Lecture Notes in Mathematics 1795

Editors:
J.-M. Morel, Cachan
F. Takens, Groningen
B. Teissier, Paris

Springer
Berlin
Heidelberg
New York
Barcelona
Hong Kong
London
Milan
Paris
Tokyo

Huishi Li

Noncommutative
Gröbner Bases and
Filtered-Graded Transfer

Springer

Author

Huishi LI

Department of Mathematics
Bilkent University
P.O. Box 217
06533 Ankara
Turkey

e-mail: huishi@fen.bilkent.edu.tr
http://www.fen.bilkent.edu.tr/ huishi

Cataloging-in-Publication Data applied for
Bibliograhpic information published by Die Deutsche Bibliothek
Die Deutsche Bibliothek lists this publication in the Deutsche Nationalbibliografie;
detailed bibliographic data is available in the Internet at http://dnb.ddb.de

Mathematics Subject Classification (2000): 16Z05, 68W30, 16W70, 16S99

ISSN 0075-8434
ISBN 3-540-44196-4 Springer-Verlag Berlin Heidelberg New York

Springer-Verlag Berlin Heidelberg New York a member of BertelsmannSpringer
Science + Business Media GmbH

http://www.springer.de

© Springer-Verlag Berlin Heidelberg 2002
Printed in Germany

Typesetting: Camera-ready TEX output by the author

SPIN: 10891005   41/3142/ du - 543210 - Printed on acid-free paper

Acknowledgement

On April 5th, 1995, on the way from Bielefeld to Xi'an, I stopped at Beijing and met my old friend Dr. Luo Yunlun (Lao Luo). Drinking Chinese green tea, Lao Luo introduced Computational Algebra to me. It was the first time I had heard of the notion of a Gröbner basis. Soon after our meeting, Lao Luo sent me some basic material on commutative computational algebra and a tutorial book introducing *Macaulay* that enabled me to start my computational algebra seminar combined with my elementary algebraic geometry course given at the Shaanxi Normal University in the spring of 1996. Without this unforgettable story, I would not have had the idea of working on the algebraic-algorithmic aspect. Thank you, Lao Luo.

I wish to express my thanks to the referees for their valuable remarks on improving the notes.

I am grateful to the LNM editors for their helpful suggestions while I was preparing the manuscript.

Huishi Li

Contents

CHAPTER VIII
Regularity and K_0-group of Quadric Solvable
Polynomial Algebras

Introduction

It is well-known that because of the successful implementation of Buchberger's algorithm on computer (1965, 1985), the commutative Gröbner basis theory has been very powerful in both pure and applied commutative mathematics. It is equally well-known that Buchberger's algorithm has been generalized to the noncommutative area via noncommutative Gröbner bases in various contexts, for instance,

- the Gröbner bases for one-sided ideals in algebras of linear partial differential operators with polynomial coefficients (or Weyl algebras) introduced in ([Gal] 1985), the Gröbner bases for one-sided ideals in the enveloping algebras of finite dimensional Lie algebras introduced in ([AL] 1988), and more generally, the Gröbner basis theory developed in ([K-RW] 1990) for a large class of non-commutative polynomial-like algebras, i.e., the class of solvable polynomial algebras,

- the Gröbner basis theory for two-sided ideals in a noncommutative free algebra developed in ([Berg] 1978) and in ([Mor1-2] 1986, 1994),

- and the Gröbner basis theory introduced in ([Gr] 1993) for two-sided ideals in an algebra which possibly has no identity but possibly has divisors of zero.

From now on in this monograph, all strings of references will be listed *alphabetically*.

Supported by several noncommutative computer algebra systems (see an incomplete list of websites at the end of this introduction for the references), noncommutative Gröbner bases have been effectively used in many contexts of pure and applied algebra, such as the algorithmic study of PDE's, the automatic proof of functional identities, the representation of algebras etc. (see [ACG1–2], [CS], [Gr], [LC1–2], [Oak1–2], [SST], [Tak1–3]).

The development and applications of noncommutative Gröbner bases shows that while noncommutativity forces algorithms to become more sophisticated and varied, it also forces a stronger interaction between algorithmic and structural methods. In particular, since Gröbner bases are not only *algorithmic* but also *structural*, effective use of noncommutative Gröbner bases often relies on effective algorithmic analysis for the algebraic structures concerned. At this point, deep structure theory is essentially needed. Typical examples are those concerning differential operators using Gröbner bases in the algebra of linear partial differential operators with polynomial coefficients (or the Weyl algebra, see [ACG1–2], [Oak1–2], [OT], [SST], [Tak1–3]), in which the D-module structure theory (in the sense of Bernstein and Kashiwara) has been the soul. But this does not mean that algorithmic methods are confined to a passive position with respect to the algebraic structures; indeed, one may observe that Gröbner basis theory proposes and realizes algorithmically a stronger concept of noetherianity in both commutative and noncommutative associative algebras (see [BW], [CLO'], [K-RW]), that certain structural properties of associative algebras are recognized via Gröbner bases (see [G-I1–3], [G-IL], [Ufn1–2]), and that a ∂-holonomic module theory stems from an algorithmic approach to the automatic proof of functional identities (see [CS], [Zei1–2]). Therefore, to develop and use a noncommutative Gröbner basis theory is to open an effective structural-algorithmic channel in noncommutative computational algebra, as the extension/contraction problem proposed to us in [CS].

As its title shows, this self-contained monograph is *not* written with the aim of developing algorithms. Instead it tries to explore the *interaction* between the structure theory of noncommutative associative algebras and the algorithmic aspect of Gröbner basis theory through several applications of *filtered-graded transfer of Gröbner bases*. Thus, if we say in the text that a certain problem (or quantity) can be solved (or computed) algorithmically, it means that, after using a certain kind of Gröbner basis, the problem (or quantity) considered can be solved (or computed in a finite number of steps). And it also means obviously that the monograph will *not* trace a history of noncommutative Gröbner basis theory, or survey others work concerning noncommutative Gröbner bases (that is far beyond the author's goal and ability). So, the author apologizes for the possible omission of some excellent references dealing with noncommutative Gröbner bases.

The content of this monograph consists of two parts.
Part I (CH.I–CH.IV): This part forms the working basis of this book, which includes some basic structural tricks in dealing with associative k-algebras, some well-known classical and modern examples of k-algebras, an introduction to Gröbner basis in associative algebra, some basic algebraic-algorithmic structures obtained in terms of Gröbner bases, and an introduction to the filtered-graded transfer of Gröbner bases in associative algebras. The Gröbner basis theory

presented is mostly based on [Gr], [K-RW] and [Mor1–2].

Part II (CH.V–CH.VIII): This part focuses on solving some structural-computational problems in quadric solvable polynomial algebras by transfering, between filtered and the associated graded structures, the relevant algorithmic data (determined in terms of Gröbner bases) and structural properties, where solvable polynomial algebra is in the sense of Kandri-Rody and Weispfenning [K-RW].

More precisely, after the preparatory CH.I, we introduce in CH.II a division algorithm and the basic theory of (left and two-sided) Gröbner basis for a (noncommutative) associative algebra that may have divisors of zero. This is realized by distinguishing the quasi-zero elements and introducing the (left) division leading monomials . Such a Gröbner basis theory is compatible with all well-known Gröbner basis theories in the literature. With such a theory, the (left) G-Noetherian structural property of an algebra (which is more stronger than the classical (left) Noetherian property) is algorithmically characterized. As a result, the notion of (left) Dickson system is quite naturally introduced in a wide context. After presenting the solvable polynomial algebra as a perfect example, in the last section of CH.II we prove the impossibility of having a Gröbner basis theory (in the sense of §3) for the algebra of linear partial differential operators with polynomial coefficients over a field of characteristic $p > 0$. We point out, however, the possibility of applying some well-defined Gröbner bases to this algebra. Starting from CH.III, we highlight the interaction between the noncommutative algebraic structure theory and the algorithmic method of using Gröbner bases. To this end, the applications of very noncommutative Gröbner bases (in the sense of [Mor1–2]) to the noncommutative structure theory given in CH.III motivate a general filtered-graded transfer of Gröbner bases in CH.IV (from which one may also see how effective the algorithmic methods are in transfering the finiteness property between algebraic structures at different levels). In CH.V–CH.VI, a double filtered-graded transfer of Gröbner bases is used to realize the computation of Gelfand-Kirillov dimension and multiplicity of finitely generated modules over a quadric solvable polynomial algebra. At this stage, after the theoretic proof is established through a structural channel, the computational aspect becomes surprisingly simple. More clearly, let L be a left ideal of a quadric solvable polynomial algebra $A = k[a_1, ..., a_n]$ and $\mathcal{G} = \{g_1, ..., g_s\}$ a left Gröbner basis for L. If the leading monomials of $\mathcal{G}$ are given by the standard monomials

$$\mathbf{LM}(g_i) = a_1^{\alpha_{i1}} a_2^{\alpha_{i2}} \cdots a_n^{\alpha_{in}}, \ i = 1, ..., s,$$

then all computations concerned can be realized by manipulating only the finite set of n-tuples of nonnegative integers

$$\alpha(\mathcal{G}) = \left\{ R_i = (\alpha_{i1}, \alpha_{i2}, ..., \alpha_{in}) \ \middle| \ i = 1, ..., s \right\}.$$

As a consequence, an elimination (of variables) lemma for quadric solvable polynomial algebras is obtained, and this lemma is used in CH.VII to formulate a

∂-holonomicity for both functions and modules with respect to a specific class of quadric solvable polynomial algebras. In addition to its interesting structural property in module theory, this ∂-holonomicity provides a way of dealing with the extension/contraction problem in the automatic proof of functional identities by using Gröbner bases in a more general context (comparing with [CS] and [Zei1]). In the final CH.VIII, a double filtered-graded transfer of data is used again to show that quadric solvable polynomial algebras are regular in the classical sense, i.e., they have finite global dimension and K_0-group $\mathbb{Z}$.

As shown in CH.II–CH.VIII, all topics discussed in this monograph are valid for many popular algebras such as the algebra of linear partial differential operators (or the Weyl algebra) and its many deformations, enveloping algebras of finite dimensional Lie algebras and many of their deformations, many quantum algebras, and some other popular operator algebras, as well as their associated (quadratic) graded structures (with respect to the standard filtration). Hence, the author hopes that the material covered in this book may be helpful for the researchers and graduate students who are interested in developing and using Gröbner bases in various contexts of mathematics.

Finally, although the aim of this monograph is not to develop algorithms related to the topics included, all examples of application are made on a strong algorithmic basis of using noncommutative Gröbner bases, i.e., all what we need in the computation involved is to have a Gröbner basis for the problem considered. Thanks to the algorithms developed in ([FG], [Mor2]) and the computer algebra systems developed at

> http://www.fmi.uni-passau.de/algebra/projects/
> http://felix.hgb-leipzig.de/
> http://www.math.vt.edu/people/green/index.html
> http://www.cis.upenn.edu/ wilf/progs.html
> http://algo.inria.fr
> http://www.math.kobe-u.ac.jp/KAN/,

those readers interested in verifying the methods introduced in this book may find available references.

CHAPTER I
Basic Structural Tricks and Examples

We assume that the reader is familiar with basic ring theory and module theory.

Throughout this book, unless otherwise stated, k will denote a (commutative) field of *arbitrary* characteristic. All rings considered are associative rings with identity 1. If A and B are rings, then $A \subset B$ means that A is a subring of B with the *same* 1. Furthermore, if $\psi \colon A \to B$ is a ring homomorphism, then we *insist* that $\psi(1) = 1$, and write $\mathrm{Ker}\psi$ and $\mathrm{Im}\psi$ for the kernel and image of ψ, respectively. We also assume that all modules considered are unitary *left* modules.

Let A be a ring and $F = \{f_i\}_{i \in J}$ a nonempty subset of A. To emphasize the *generators* of the *two-sided* ideal generated by F in a formula, we sometimes use $\langle f_i \rangle_{i \in J}$ to indicate this ideal without making confusion, and similarly write $\langle f_i]$ for the *left* ideal of A generated by F.

Conventionally, we use $I\!N$, $Z\!\!\!Z$, $Q\!\!\!Q$, $I\!R$, and $C\!\!\!C$ for the sets of non-negative integers, integers, rational numbers, real numbers, and complex numbers, respectively.

1. Algebras and their Defining Relations

Since both commutative and noncommutative Gröbner bases are constructed on the existence of a division algorithm, while a division algorithm essentially manipulates "ordered bases" of algebras which are determined by their defining relations, we start this   preparatory chapter by recalling   some basic elements concerning algebras.

Let k be a field. A *k-algebra* is a ring A together with a *nonzero* ring homomorphism $\eta_A\colon k \to A$ satisfying

(**A1**) the image of η_A, denoted $\operatorname{Im}\eta_A$, is contained in the center of A, and
(**A2**) the map

$$\begin{aligned} k \times A &\longrightarrow A \\ (\lambda, a) &\longmapsto \eta_A(\lambda)a \end{aligned}$$

equips A with a vector space structure over k and the multiplicative map $\mu_A\colon A \times A \to A$ is bilinear:

$$\begin{aligned} \mu_A(a+b,c) &= \mu_A(a,c) + \mu_A(b,c), & a,b,c &\in A, \\ \mu_A(a,b+c) &= \mu_A(a,b) + \mu_A(a,c), & a,b,c &\in A, \\ \mu_A(\lambda a, b) &= \mu_A(a, \lambda b) = \lambda\mu(a,b), & \lambda &\in k,\ a,b \in A. \end{aligned}$$

If A is a commutative ring, we say that A is a *commutative k-algebra*.

It is clear that any A-module M also has a k-vector space structure. If M is a k-vector space, then the set of all linear operators on M, denoted $\operatorname{End}_k M$, forms a k-algebra with respect to the operator composite and the scalar multiplication of operators, which is known as the *k-algebra of linear operators* of M.

Let A and B be k-algebras. A *k-algebra homomorphism* from A to B is a ring homomorphism $\alpha\colon A \to B$ such that $\alpha \circ \eta_A = \eta_B$, i.e., we have the following commutative diagram:

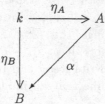

(This also means that α is a linear transformation form the k-vector space A to the k-vector space B.)

If $\ell\colon A \to B$ is an injective algebra homomorphism, we say that A is a *subalgebra* of B. (Compare this with the convention we made for subring in the beginning of this chapter.)

Remark Ignoring the conventions we made for ring homomorphisms and for subrings, from the definitions given above we may also derive the following facts:
(i) η_A is nonzero. Hence η_A is injective, $k \cong \operatorname{Im}\eta_A$ and $\eta_A(1) = 1$.
(ii) k may be viewed as a subalgebra of A, and we may write λa for $\eta_A(\lambda)a$ (the scalar multiplication of the k-vector space A).
(iii) α preserves the units, i.e., $\alpha(1) = 1$.

Let A be a k-algebra and I a two-sided ideal of A. Then the quotient ring (or factor ring) A/I of A has the k-algebra structure induced by A, called the *quotient algebra* (or *factor algebra*) of A, and the natural map $\pi\colon A \to A/I$ is a k-algebra homomorphism.

Let $X = \{X_i\}_{i \in \Lambda}$ be a set and $S = \langle X \rangle$ the (multiplicative) free semigroup on X. Then S consists of all words in the alphabet X plus the empty word $\emptyset$, i.e.,

$$S = \left\{ X_{i_1} X_{i_2} \cdots X_{i_p} \mid X_{i_j} \in X,\ p \geq 1 \right\} \cup \{\emptyset\},$$

and the multiplication on S is defined as the *concatenation* of words, that is, if $u = X_{i_1} \cdots X_{i_p}$ and $w = X_{i_{p+1}} \cdots X_{i_n}$ then

$(*)$
$$\begin{aligned} &\emptyset u = u\emptyset = u,\ \text{and} \\ &uw = X_{i_1} \cdots X_{i_p} X_{i_{p+1}} \cdots X_{i_n} \end{aligned}$$

Consider the k-vector space $k\langle X \rangle$ with basis the set S. Then formula $(*)$ equips $k\langle X \rangle$ with an algebra structure, called the *free k-algebra* on the set X. If $X = \{X_1, ..., X_n\}$ is finite, we also denote $k\langle X \rangle$ by $k\langle X_1, ..., X_n \rangle$.

Free k-algebras have the following universal property.

1.1. Theorem Let $X = \{X_i\}_{i \in \Lambda}$. Given a k-algebra A and a set-theoretic map $\varphi\colon X \to A$, there exists a unique algebra homomorphism $\overline{\varphi}\colon k\langle X \rangle \to A$ such that $\overline{\varphi}(X_i) = \varphi(X_i)$ for all $X_i \in X$, i.e., we have the following commutative diagram:

where ℓ is the inclusion map.

Proof It is sufficient to define $\overline{\varphi}$ on any word of S. For the empty word we set $\overline{\varphi}(\emptyset) = 1$. Otherwise, if $u = X_{i_1} \cdots X_{i_p} \in S$, we define

$$\overline{\varphi}(u) = \varphi(X_{i_1}) \cdots \varphi(X_{i_p}).$$

The remained proof follows easily. $\qquad\qquad\square$

Let A be a k-algebra and $T = \{a_i\}_{i \in \Lambda} \subset A$ a nonempty subset of A. Write $k[T]$ for the subalgebra of A generated by T and k, i.e.,

$$k[T] = \left\{ \sum \lambda_{i_1 i_2 \cdots i_s} a_{i_1} a_{i_2} \cdots a_{i_s} \ \middle|\ \lambda_{i_1 i_2 \cdots i_s} \in k,\ a_{i_j} \in T,\ s \in I\!N \right\}.$$

If $k[T] = A$, we say that A *is generated by* T and that T is a *generating set* of A (or a set of *generators* of A). If $T = \{a_1, ..., a_n\}$ is finite and $A = k[T]$, we say that A is a *finitely generated* k-algebra and write $A = k[a_1, a_2, ..., a_n]$.

Since every k-algebra A has a generating set T (for instance $T = A$), say $T = \{a_i\}_{i \in \Lambda}$, if we take the set of symbols $X = \{X_i\}_{i \in \Lambda}$ and consider the free algebra $k\langle X \rangle$ on X and the set-theoretic map $\varphi \colon X \to T$ with $\varphi(X_i) = a_i$, it follows from Theorem 1.1 that there is a unique k-algebra homomorphism $\overline{\varphi} \colon k\langle X \rangle \to A = k[T]$, which is clearly onto, such that

$$(**) \qquad\qquad\qquad A \cong k\langle X \rangle / I$$

where $I = \mathrm{Ker}\,\overline{\varphi}$. Thus, the following holds.

1.2. Proposition Any k-algebra $A = k[T]$ with $T = \{a_i\}_{i \in \Lambda}$ is the quotient of a free k-algebra $k\langle X \rangle$ on a set X.

$\square$

1.3. Definition In the above $(**)$, if the two-sided ideal I of $k\langle X \rangle$ is generated by

$$\mathcal{F} = \left\{ f_j = \sum \lambda_{ij} X_{1j} X_{2j} \cdots X_{sj} \;\middle|\; j \in J \right\},$$

we say that $\mathcal{F}$ is a set of *defining relations* of the k-algebra A. In this case we also say that the k-algebra A is generated by $T = \{a_i\}_{i \in \Lambda}$ subject to the relations:

$$f_j(a_{1j} a_{2j} \cdots a_{sj}) = \sum \lambda_{ij} a_{1j} a_{2j} \cdots a_{sj} = 0, \quad j \in J.$$

Example (i) As the first example, let $X = \{X_i\}_{i \in \Lambda}$ and $k\langle X \rangle$ be as before. Let I be the two-sided ideal of $k\langle X \rangle$ generated by all elements of the form

$$X_j X_i - X_i X_j, \quad j,\, i \in \Lambda,\, j \neq i.$$

Then the algebra $A = k\langle X \rangle / I = k[x_i]_{i \in \Lambda}$ is a commutative algebra generated by $\{x_i\}_{i \in \Lambda}$, where each x_i denotes the image of X_i in A, subject to the relations:

$$x_j x_i = x_i x_j, \quad j,\, i \in \Lambda,\, j \neq i,$$

and is called the *symmetric algebra* on X. Using Theorem 1.1 and the fundamental theorem of homomorphism, one may easily show that A has the following universal property:

- Given a commutative algebra B and a set-theoretic map $\varphi \colon \{x_i\}_{i \in \Lambda} \to B$, then there exists a unique algebra homomorphism $\overline{\varphi} \colon A \to B$ such that $\overline{\varphi}(x_i) = \varphi(x_i)$ for all $x_i \in \{x_i\}_{i \in \Lambda}$, i.e., we have the following commutative diagram:

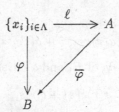

Hence, every commutative algebra is the quotient of a symmetric algebra.
If $X = \{X_1, ..., X_n\}$ is finite, then $A = k[x_1, ..., x_n]$ is nothing but the commutative polynomial k-algebra in n variables. And every finitely generated commutative algebra is the quotient of a polynomial algebra $k[t_1, t_2, ..., t_n]$ in variables $t_1, ..., t_n$ for some n.

2. Skew Polynomial Rings

An important class of noncommutative algebras, for which a Gröbner basis theory may exist, is the class of skew polynomial algebras. We review the basic structure of a skew polynomial ring (algebra) by following [MR].

Skew polynomial rings are polynomial rings in noncommutative setting. More precisely, we want to have polynomials over a ring R in a variable x, $\sum_{i=1}^{n} r_i x^i$, *which is not assumed to commute with the elements of R*, but is desired that

(S1) each polynomial should be expressed uniquely in the form $\sum r_i x^i$ for some $r_i \in R$, and

(S2) $xr \in Rx + R$, i.e., $xr = \sigma(r)x + \delta(r)$ for some $\sigma(r), \delta(r) \in R$.

The *degree* of a polynomial $f(x) = \sum_{i=1}^{n} r_i x^i$ is defined to be n provided $r_n \neq 0$. If $r_n \neq 0$ we call r_n the *leading coefficient* of $f(x)$ and write $\deg f(x) = n$. If all the coefficients r_i of $f(x)$ are zero, we say that $f(x)$ is the zero polynomial and write $f(x) = 0$. Conventionally the zero polynomial has degree $-\infty$ and leading coefficient 0.

Thus, if we apply the condition **(S1)** to xr, $r \in R$, then the condition **(S2)** guarantees that the degrees behave appropriately:

$$
(*) \quad \begin{cases}
\deg f(x) \geq 0, \\
\deg f(x) = -\infty \text{ if and only if } f(x) = 0, \\
\deg(f(x) - g(x)) \leq \max\{\deg f(x), \deg g(x)\}, \\
\deg(f(x)g(x)) \leq \deg f(x) + \deg g(x).
\end{cases}
$$

And under these conditions, it is clear that σ, δ are $\mathbb{Z}$-*operators* on the additive group of R, ie., $\sigma, \delta \in \text{End}_{\mathbb{Z}} R$, the ring of automorphisms of the additive group of R. Moreover, for $r,\ s \in R$,

$$x(rs) = \sigma(rs)x + \delta(rs)$$
$$(xr)s = \sigma(r)\sigma(s)x + \sigma(r)\delta(s) + \delta(r)s.$$

Thus σ is a *ring endomorphism* of R, and δ satisfies

$(**)$ $\qquad\qquad\qquad \delta(rs) = \sigma(r)\delta(s) + \delta(r)s, \quad r, s \in R.$

Note in particular that $\sigma(1) = 1$ and $\delta(1) = 0$.

2.1. Definition Let σ be a ring endomorphism of R and $\delta \in \text{End}_{\mathbb{Z}} R$. δ is said to be a *σ-derivation* if it satisfies the above $(**)$.

Conversely, let R be a ring, σ a ring endomorphism of R and δ a σ-derivation. We now construct a noncommutative polynomial ring over R which has exactly the properties **(S1)–(S2)** listed above.

Let $E = \text{End}_{\mathbb{Z}} R^{\mathbb{N}}$ be the ring of $\mathbb{Z}$-operators on $R^{\mathbb{N}}$, where $R^{\mathbb{N}} = \prod_{i \in \mathbb{N}} R_i$ is the direct product of the additive groups $R_i = R$. Then $R \hookrightarrow E$, acting by left multiplication. Define

$$x : \quad R^{\mathbb{N}} \quad \longrightarrow \quad R^{\mathbb{N}}$$

$$(r_i) \quad \mapsto \quad (\sigma(r_{i-1}) + \delta(r_i)),$$

where $(r_i) = (r_0, r_1, ...) \in R^{\mathbb{N}}$ and $r_{-1} = 0$.
Let $R[x; \sigma, \delta]$ denote the subring of E generated by R and x. Then, viewing $r \in R$ as $\mathbb{Z}$-operator on $R^{\mathbb{N}}$, we have

$$xr = \sigma(r)x + \delta(r), \quad r \in R,$$

and consequently $f(x) = \sum r_i x^i$ for any $f(x) \in R[x; \sigma, \delta]$. Since

$$\sum r_i x^i (1, 0, 0, ...) = (r_i),$$

it follows that the expression for each $f(x) \in R[x; \sigma, \delta]$ is unique. Thus we have constructed the noncommutative polynomial ring with the desired properties.

2.2. Definition The ring $R[x; \sigma, \delta]$ is called the *skew polynomial ring* over R.

Example (i) Skew polynomial algebras on $k[x]$.
Let $k[x]$ be the polynomial algebra over a field k in one variable x, and let σ be a k-algebra endomorphism of $k[x]$. Suppose $\sigma(x) = g(x)$, and that δ is a σ-derivation on $k[x]$ with $\delta(x) = f(x)$. If we construct the skew polynomial ring $k[x][y; \sigma, \delta]$, then

$$yx = \sigma(x)y + \delta(x) = g(x)y + f(x).$$

It follows that

$$\delta(x^i) \quad = \quad g(x)^{i-1}f(x) + g(x)^{i-2}f(x)x + g(x)^{i-3}f(x)x^2 + \cdots$$
$$+ g(x)f(x)x^{i-2} + f(x)x^{i-1}.$$

Conversely, given any $f(x) \in k[x]$ we can define a σ-derivation as above and such that $\delta(x) = f(x)$.

Now the following is clear:

(a) If $\sigma(x) = g(x) = x$ is the identity endomorphism of $k[x]$, then $\delta = f(x)\frac{\partial}{\partial x}$.

(b) $k[x][y; \sigma, \delta] \cong \frac{k\langle X, Y\rangle}{\langle YX - g(X)Y - f(X)\rangle}$, where $k\langle X, Y\rangle$ is the free algebra generated by X and Y, i.e., $k[x][y; \sigma, \delta]$ is the k-algebra generated by two elements x and y subject to the relation:

$$yx = g(x)y + f(x).$$

Remark (i) In the literature $R[x; \sigma, \delta]$ is also called an Ore extension of R because of the remarkable historical literature [Ore].

(ii) $R[x; \sigma, \delta]$ can also be described as being the ring T generated freely over R by an element X, as follows. Consider the free R-module $T = \oplus_{i=0}^{\infty} RX^i$ and, then define the multiplication on T by putting $Xr = \sigma(r)X + \delta(r)$, $X^iX^j = X^{i+j}$. Thus there is an obvious onto ring homomorphism $T \to R[x; \sigma, \delta]$. Since the x^i are R-independent in $R[x; \sigma, \delta]$, they are also R independent in T. Hence $T \cong R[x; \sigma, \delta]$.

From the definition it is clear that if $\sigma = 1$ and $\delta = 0$ then $R[x; \sigma, \delta]$ is nothing but the usual commutative polynomial ring over R in one variable x. Moreover, we have the following easily verified facts:

a. There is an obvious embedding of R into $R[x; \sigma, \delta]$ with $r \mapsto rx^0$.

b. $R[x; \sigma, \delta]/xR[x; \sigma, \delta] \cong R$ as left R-modules. If $\delta = 0$, then $xR[x; \sigma, \delta]$ becomes a two-sided ideal of $R[x; \sigma]$ and the foregoing isomorphism becomes a ring isomorphism.

c. $R[x; \sigma, \delta]/(x-1)R[x; \sigma, \delta] \cong R$ as left R-modules.

Next, we list some basic properties of $R[x; \sigma, \delta]$, and we refer to ([MR] CH.I §§2, 4, [Kass] CH.I 1.7) for the detailed proofs.

2.3. Theorem (universal property) If $\psi \colon R \to S$ is a ring homomorphism and $y \in S$ has the property that

$$y\psi(r) = \psi(\sigma(r))y + \psi(\delta(r)), \quad \text{for all } r \in R,$$

then there exists a unique ring homomorphism $\chi \colon R[x; \sigma, \delta] \to S$ such that

$\chi(x) = y$ and the following diagram commutes:

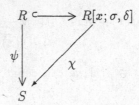

As before, we say that the polynomial $f(x) = \sum_{i=1}^{n} r_i x^i$ has *degree n* provided $r_n \neq 0$, denoted $\deg f(x) = n$; and then r_n is called the *leading coefficient* of $f(x)$. If $f(x) = 0$ we put $\deg f(x) = -\infty$ and we say that $f(x)$ has leading coefficient 0. Thus the *degree function*

$$d \; : \; R[x;\sigma,\delta] \; \longrightarrow \; N \cup \{-\infty\}$$
$$f(x) \; \longmapsto \; \deg f(x)$$

is well defined on $R[x;\sigma,\delta]$ in the sense of the foregoing $(*)$.

2.4. Proposition If $R = k$ is a field, then $k[x;\sigma,\delta]$ satisfies the (left) Euclidean division algorithm relative to the degree function d:

- for $f, g \in k[x;\sigma,\delta]$ with $g \neq 0$, there exists $q, r \in k[x;\sigma,\delta]$ such that $f = qg + r$, $d(r) < d(g)$.

It follows that in this case $k[x;\sigma,\delta]$ is a principal left ideal ring.

□

2.5. Proposition Let $A = R[x;\sigma,\delta]$.
(i) If σ is injective and R is a domain (i.e., R does not have nontrivial divisors of zero), then A is a domain.
(ii) If σ is injective and R is a division ring (i.e., R is a noncommutative field), then A is a principal left ideal domain.
(iii) If σ is a ring automorphism of R, and R is a prime ring, then A is a prime ring.
(iv) If σ is a ring automorphism of R and R is a left (right) Noetherian ring, then A is a left (right) Noetherian ring.

□

Note that, for $r \in R$,

$$x^n r = \alpha_n(r)x^n + \alpha_{n-1}(r)x^{n-1} + \cdots + \alpha_0(r),$$

where each $\alpha_i \in \text{End}_{\mathbb{Z}} R$. So if $\alpha_n(r) \neq 0$ then $x^n r$ has degree n and leading coefficient $\alpha_n(r)$. The general formula for α_i in terms of σ, δ and n are   rather complicated, but for $R[x;\sigma]$ and $R[x;\delta]$ they are simple. For $R[x;\sigma]$, $\alpha_n = \sigma^n$ and $\alpha_i = 0$ for $i \neq n$; and for $R[x;\delta]$,

$$\alpha_i = \binom{n}{i} \delta^{n-i}.$$

(Compare Leibnitz's formula for $\delta^n(rs)$: $(rs)^{(n)} = \sum_{i=0}^{n} \binom{n}{i} r^{(i)} s^{(n-i)}$.)

In the case where σ is injective, it is possible to give a general formula for the product in $R[x;\sigma,\delta]$. Consider $P = \sum_{i=0}^{n} r_i x^i$ and $Q = \sum_{i=0}^{m} s_i x^i$. Set $PQ = \sum_{i=0}^{n|m} u_i x^i$. Let $S_{n,k}$ be the element in $\mathrm{End}_{\mathbb{Z}} R$ defined as the sum of all $\binom{n}{k}$ possible compositions of k copies of δ and of $n-k$ copies of σ.

2.6. Proposition With notation as above, if σ is injective, then for all i with $0 \leq i \leq m+n$ we have

$$u_i = \sum_{p=0}^{i} r_p \sum_{k=0}^{p} S_{p,k}(s_{i-p+k})$$

and for all $r \in R$ and $n \in \mathbb{N}$ we have in $R[x;\sigma,\delta]$

$$x^n r = \sum_{k=0}^{n} S_{n,k}(r) x^{n-k}.$$

$\square$

3. $\mathbb{Z}$-filtrations and their Associated Graded Structures

This section and the next are devoted to the notions and basic tricks concerning $\mathbb{Z}$-filtrations and their associated graded structures that will be used to develop the later chapters. We refer to [NVO] and [LVO4] for a general theory on $\mathbb{Z}$-filtered and $\mathbb{Z}$-graded rings. The $\succeq_{gr}$-filtration on a quadric solvable polynomial algebra is introduced and discussed in CH.V and CH.VIII.

First look at graded rings.

3.1. Definition A ring A is said to be $\mathbb{Z}$-graded if $A = \oplus_{p \in \mathbb{Z}} A_p$, where the A_p are subgroups of the additive group of A, such that

$$A_p A_q \subset A_{p+q} \quad \text{for all } p, q \in \mathbb{Z}.$$

For $p \in \mathbb{Z}$, A_p is called the pth *homogeneous part* (or the *part of degree p*) of A. An element $a_p \in A_p$ is called a *homogeneous element of degree p*.

If $A_p = 0$ for all $p < 0$, i.e., $A = \oplus_{p \in \mathbb{N}} A_p$, then A is called a *positively graded ring*.

If A is a k-algebra over a field k, we require that each A_p is a subspace of the k-vector space A.

Note that if $A = \oplus_{\mathbb{Z}} A_p$ is a $\mathbb{Z}$-graded ring (k-algebra), then it follows from the definition that A_0 is automatically a subring (subalgebra) of A, and $1_A \in A_0$ is an element of degree 0.

3.2. Definition Let A be a $\mathbb{Z}$-graded ring. An A-module M is said to be a *graded A-module* if $M = \oplus_{p \in \mathbb{Z}} M_p$, where the M_p are subgroups of the additive group of M, such that

$$A_p M_q \subset M_{p+q} \quad \text{for all } p, q \in \mathbb{Z}.$$

For $p \in \mathbb{Z}$, M_p is called the pth *homogeneous part* (or the *part of degree p*) of M. An element $m_p \in M_p$ is called a *homogeneous element of degree p*.

If A is a k-algebra over a field k, we usually require that each M_p is a subspace of the k-vector space M.

If A is a $\mathbb{Z}$-graded ring and M is a graded A-module, then it follows from Definition 3.1–3.2 that every element $m \in M$ has a unique expression as a sum of homogeneous elements:

$$m = m_p + m_{p-1} + \cdots + m_0, \quad m_i \in M_i.$$

3.3. Proposition Let M be a graded A-module and N a submodule of M. Then the following are equivalent.

(i) $N = \oplus_{p \in \mathbb{Z}} (M_p \cap N)$,

(ii) If $u \in N$, then all homogeneous components of u are contained in N.

(iii) N is generated by homogeneous elements.

(iv) The quotient module $M/N = \sum_{p \in \mathbb{Z}} (M_p + N)/N$ is graded as $M/N = \oplus_{p \in \mathbb{Z}} (M/N)_p$ with $(M/N)_p = (M_p + N)/N$.

$\square$

3.4. Definition Let A be a graded ring and M a graded A-module. A submodule N of M satisfying one of the equivalent conditions of Proposition 3.3 is called a *graded submodule* of M.

If submodules are replaced by left (two-sided) ideals of A in Proposition 3.3, we arrive at the definition of a *graded left ideal* (*graded ideal*).

3.5. Definition (i) Let $A = \oplus_{p \in \mathbb{Z}} A_p$ and $B = \oplus_{p \in \mathbb{Z}} B_p$ be graded rings (graded algebras). A ring homomorphism (algebra homomorphism) $\varphi \colon A \to B$ is called a *graded ring homomorphism* (*graded algebra homomorphism*) if $\varphi(A_p) \subset B_p$ for all $p \in \mathbb{Z}$.
(ii) Let $M = \oplus_{p \in \mathbb{Z}} M_p$ and $N = \oplus_{p \in \mathbb{Z}} N_p$ be graded modules over a graded ring (graded algebra) $A = \oplus_{p \in \mathbb{Z}} A_p$. An A-homomorphism $\varphi \colon M \to N$ is called a *graded A-homomorphism* if $\varphi(M_p) \subset N_p$ for all $p \in \mathbb{Z}$.

As a consequence of Proposition 3.3 and the above definitions, the following holds.

3.6. Corollary (i) If $\varphi \colon \oplus_{p \in \mathbb{Z}} A_p = A \to B = \oplus_{p \in \mathbb{Z}} B_p$ is a graded ring homomorphism, then $\mathrm{Im}\varphi$ is a *graded subring* of B in the sense that $\mathrm{Im}\varphi = \oplus_{p \in \mathbb{Z}}(B_p \cap \mathrm{Im}\varphi)$, and $\mathrm{Ker}\varphi$ is a graded ideal of A.
(ii) If $\psi \colon \oplus_{p \in \mathbb{Z}} M_p = M \to N = \oplus_{p \in \mathbb{Z}} N_p$ is a graded A-homomorphism, then $\mathrm{Im}\psi$ is a graded submodule of N and $\mathrm{Ker}\psi$ is a graded submodule of M.

$\square$

Example (i) Let $k\langle X \rangle$ be the free k-algebra on $X = \{X_i\}_{i \in \Lambda}$, and let S be the free semigroup generated by X (see §2). If $w = X_{i_1} X_{i_2} \cdots X_{i_p}$ is a word of S, then the length p of w is called the *degree* of w and is denoted by $d(w)$. We write $d(1) = 0$. For each $p \in \mathbb{N}$, consider the k-subspace of $k\langle X \rangle$:

$$k\langle X \rangle_p = \left\{ \sum_{d(w)=p} c_w w \;\middle|\; c_w \in k,\ w \in S \right\}.$$

Then $k\langle X \rangle = \oplus_{p \in \mathbb{N}} k\langle X \rangle_p$ is a positively graded k-algebra with $k\langle X \rangle_0 = k$.
If I is an ideal of $k\langle X \rangle$ generated by homogeneous elements, then by Proposition 3.3, the k-algebra $A = k\langle X \rangle/I$ is a positively graded algebra, i.e., $A = \oplus_{p \in \mathbb{N}} A_p$ with $A_p = (k\langle X \rangle_p + I)/I$. The gradation given on A in such a way is called the *natural gradation* on A. For instance, the commutative polynomial algebra $A = k[x_1, ..., x_n]$ in n variables has the natural gradation

$$A_p = \left\{ \sum \lambda_\alpha x_1^{\alpha_1} \cdots x_n^{\alpha_n} \;\middle|\; \lambda_\alpha \in k,\ \alpha_1 + \cdots + \alpha_n = p \right\}, \quad p \in \mathbb{N},$$

i.e., each A_p consists of all homogeneous polynomial of degree p.

(ii) Let $A = \oplus_{p \in \mathbb{Z}} A_p$ be a graded ring. Consider the polynomial ring $A[t]$ over A in one commuting variable t. Then $A[t]$ may be graded by the *mixed gradation*: $A[t] = \oplus_{p \in \mathbb{Z}} A[t]_p$, where

$$A[t]_p = \left\{ \sum_{i+j=p} a_i t^j \;\middle|\; a_i \in A_i,\ j \in \mathbb{N} \right\}, \quad p \in \mathbb{Z}.$$

Now let us turn to filtered rings.

3.7. Definition A ring A is called a $\mathbb{Z}$-*filtered ring* with a filtration FA if there is a sequence $FA = \{F_pA\}_{p \in \mathbb{Z}}$ of subgroups of the additive group of A such that

(F1) $\displaystyle \bigcup_{p \in \mathbb{Z}} F_pA = A$,

(F2) $1 \in F_0A$,

(F3) $F_{p-1}A \subset F_pA$ for all $p \in \mathbb{Z}$,

(F4) $(F_pA)(F_qA) \subset F_{p+q}A$ for all $p, q \in \mathbb{Z}$,

We say that FA is *separated* if it also satisfies:

(F5) $\displaystyle \bigcap_{p \in \mathbb{Z}} F_pA = \{0\}$.

If $F_{-1}A = 0$ then A is called a *positively filtered* ring and FA is called a *positive filtration* on A. Obviously positive filtrations are separated.

If A is a k-algebra over a field k, we usually require that each F_pA is a subspace of the k-vector space A.

Let A be a $\mathbb{Z}$-filtered ring with filtration FA. The *associated graded ring* of A with respect to FA, denoted $G(A)$, is defined to be the $\mathbb{Z}$-graded ring $G(A) = \oplus_{p \in \mathbb{Z}} G(A)_p$ with $G(A)_p = F_pA/F_{p-1}A$ and the multiplication

$$\frac{F_pA}{F_{p-1}A} \times \frac{F_qA}{F_{q-1}A} \qquad \longrightarrow \qquad \frac{F_{p+q}A}{F_{p+q-1}A}$$

$$(a + F_{p-1}A,\ b + F_{q-1}A) \qquad \longmapsto \qquad ab + F_{p+q-1}A$$

The *Rees ring* of A with respect to FA, denoted $\widetilde{A}$, is defined to be the $\mathbb{Z}$-graded ring $\widetilde{A} = \oplus_{p \in \mathbb{Z}} \widetilde{A}_p$ with $\widetilde{A}_p = F_pA$ and the multiplication

$$\widetilde{A}_p \times \widetilde{A}_q \qquad \longrightarrow \qquad \widetilde{A}_{p+q}$$

$$(h(a)_p,\ h(b)_q) \qquad \longmapsto \qquad h(ab)_{p+q}$$

where $h(a)_p$ denotes the homogeneous element in $\widetilde{A}_p$ represented by $a \in F_nA$, $n \leq p$.

Let X denote the homogeneous element of degree 1 in $\widetilde{A}_1 = F_1A$ represented by 1. To be convenient, we call X the *canonical element* of $\widetilde{A}$.

3.8. Proposition With notation as above, the following holds.

(i) The canonical element X is contained in the center of $\widetilde{A}$ and is not a divisor of zero. Hence we have the two-sided ideals $\langle X \rangle = X\widetilde{A}$, $\langle 1 - X \rangle = (1 - X)\widetilde{A}$. Moreover, $(1 - X)\widetilde{A}$ does not contain any nonzero homogeneous element of $\widetilde{A}$.

(ii) $A \cong \widetilde{A}/(1 - X)\widetilde{A}$.

(iii) $G(A) \cong \widetilde{A}/X\widetilde{A} = \oplus_{p \in \mathbb{Z}}(\widetilde{A}_p + X\widetilde{A})/X\widetilde{A}$ as graded rings.

Proof (i) This may be directly verified.

(ii) Note that if $h(a_i)_{p_i} \in \widetilde{A}_{p_i}$ is the homogeneous element of degree p_i represented by $a_i \in F_n A$, $n \leq p_i$, and if $\ell \geq p_i$, then $X^{\ell - p_i} h(a_i)_{p_i} = h(a_i)_\ell \in \widetilde{A}_\ell$ is the homogeneous element of degree ℓ represented by $a_i \in F_n A$. Thus, if $\sum h(a_i)_{p_i} \in \widetilde{A}$ with $h(a_i)_{p_i} \in \widetilde{A}_{p_i}$ and $\ell \geq p_i$, then

$$\sum h(a_i)_{p_i} = \sum \left(X^{\ell - p_i} h(a_i)_{p_i} - X^{\ell - p_i} h(a_i)_{p_i} + h(a_i)_{p_i} \right)$$
$$= h\left(\sum a_i \right)_\ell + \sum h(a_i)_{p_i} \left(1 - X^{\ell - p_i} \right).$$

Now consider the ring homomorphism

$$\psi: \quad \widetilde{A} \quad \longrightarrow \quad A$$

$$\sum h(a_i)_{p_i} \quad \mapsto \quad \sum a_i$$

Then it is easy to see that ψ is surjective and $\mathrm{Ker}\psi = (1 - X)\widetilde{A}$. Hence (ii) is proved.

(iii) Note that since $G(A)_p = F_p A/F_{p-1} A$ and $(\widetilde{A}/X\widetilde{A})_p = (\widetilde{A}_p + X\widetilde{A})/X\widetilde{A}$ with $\widetilde{A}_p = F_p A$. It is not hard to construct the desired graded ring isomorphism by using the formula $Xh(a)_{p-1} = h(a)_p$ for $a \in F_{p-1}A$. $\square$

Another way to understand the Rees ring $\widetilde{A}$ is to identify it with the *graded subring* $\overline{A} = \oplus_{n \in \mathbb{Z}} F_n A t^n$ of $A[t, t^{-1}]$, where $A[t, t^{-1}]$ is the ring of Laurent polynomials (i.e., the formal polynomials $\sum a_i t^i$ $a_i \in A$, $i \in \mathbb{Z}$) in the commuting variable t, and the addition and multiplication on $A[t, t^{-1}]$ are defined as in the usual polynomial ring. The $\mathbb{Z}$-gradation on $A[t, t^{-1}]$ is given by the degree of polynomials, and the graded ring isomorphism from $\widetilde{A}$ to $\overline{A}$ is defined as:

$$\bigoplus_{n \in \mathbb{Z}} F_n A = \widetilde{A} \quad \longrightarrow \quad \overline{A} = \bigoplus_{n \in \mathbb{Z}} F_n A t^n$$

$$\sum h(a_i)_{p_i} \quad \mapsto \quad \sum a_i t^{p_i}$$

Obviously, t is a homogeneous central element of degree 1 in $\overline{A}$ which is also not a divisor of zero, and $X \mapsto t$. (This may be easily recaptured from Proposition 3.8.)

The Rees ring $\widetilde{A}$ of a filtered ring A constructed with respect to a given filtration FA has been widely used in the literature, e.g., [AVO], [Gin], [Kass] and [LVO4]. Also see CH.III §3 for some exposition.

3.9. Definition Let A be a $\mathbb{Z}$-filtered ring with filtration FA, and M an A-module. If there is a sequence $FM = \{F_pM\}_{p \in \mathbb{Z}}$ of subgroups of the additive group of M such that

(FM1) $\bigcup_{p \in \mathbb{Z}} F_pM = M$,

(FM2) $F_{p-1}M \subset F_pM$ for all $p \in \mathbb{Z}$,

(FM3) $(F_pA)(F_qM) \subset F_{p+q}M$ for all $p, q \in \mathbb{Z}$,

then M is called a *filtered A-module* with filtration FM. If $F_nM = 0$ for some $n \in \mathbb{Z}$, then we say that FM is a *discrete filtration* on M. (Obviously the positive filtration on a filtered ring is discrete.) We say that FM is *separated* if it also satisfies:

(FM4) $\bigcap_{p \in \mathbb{Z}} F_pM = \{0\}$.

If A is a k-algebra over a field k, we require that each F_pM is a subspace of the k-vector space M.

Let M be a filtered A-module with filtration FM. The *associated graded module* of M with respect to FM, denoted $G(M)$, is defined to be the graded $G(A)$-module $G(M) = \oplus_{p \in \mathbb{Z}} G(M)_p$ with $G(M)_p = F_pM/F_{p-1}M$, where the module action of $G(A)$ on $G(M)$ is given by

$$\frac{F_pA}{F_{p-1}A} \times \frac{F_qM}{F_{q-1}M} \qquad \longrightarrow \qquad \frac{F_{p+q}M}{F_{p+q-1}M}$$

$$(a + F_{p-1}A, \; m + F_{q-1}M) \quad \mapsto \quad am + F_{p+q-1}M$$

The *Rees module* of M with respect to FM, denoted $\widetilde{M}$, is defined to be the graded $\widetilde{A}$-module $\widetilde{M} = \oplus_{p \in \mathbb{Z}} \widetilde{M}_p$ with $\widetilde{M}_p = F_pM$ and the module action

$$\widetilde{A}_p \times \widetilde{M}_q \qquad \longrightarrow \qquad \widetilde{M}_{p+q}$$

$$(h(a)_p, \; h(m)_q) \quad \mapsto \quad h(am)_{p+q}$$

where $h(a)_p$ denotes the homogeneous element in $\widetilde{A}_p$ represented by $a \in F_nA$ with $n \le p$, and $h(m)_q$ denotes the homogeneous element in $\widetilde{M}_q$ represented by $m \in F_sM$ with $s \le q$.

Relations between $\widetilde{M}$, M, and $G(M)$ are mentioned as follows.

3.10. Proposition With notation as above, let X be the canonical element of $\widetilde{A}$. Then the following holds.

(i) X does not annihilate any nonzero element of $\widetilde{M}$, and $(1 - X)\widetilde{M}$ does not contain any nonzero homogeneous element of $\widetilde{M}$.

(ii) $M \cong \widetilde{M}/(1-X)\widetilde{M}$.
(iii) $G(M) \cong \widetilde{M}/X\widetilde{M}$.

□

Any filtration FM on an A-module M defines an *order function* $v\colon M \to \mathbb{Z}$ as follows:

$$v(m) = \begin{cases} -\infty, & \text{if } m \in \cap_{n \in \mathbb{Z}} F_n M, \\ \\ p, & \text{if } m \in F_p M - F_{p-1} M. \end{cases}$$

Important notation If $m \in M$ with $v(m) = p$, we write $\sigma(m)$ for the *nonzero* image of m in $G(M)_p = F_p M/F_{p-1}M$, and write $\widetilde{m}$ for the homogeneous element $h(m)_p$ in $\widetilde{M}_p$.

3.11. Lemma With notation as before, the following holds.
(i) If $m \notin \cap_{n \in \mathbb{Z}} F_n M$, then any homogeneous element in $\widetilde{M}$ represented by m can be uniquely written as

$$X^s \widetilde{m}$$

for some $s \geq 0$.
(ii) If $a \in A$ with $v(a) = p$, and $m \in M$ with $v(m) = q$ such that $am \notin \cap_{n \in \mathbb{Z}} F_n M$, then

$$\widetilde{a}\widetilde{m} = X^s \widetilde{am},$$

where $s = p + q - v(am)$.
(iii) Let $a \in A$, $m \in M$. Then $\sigma(a)\sigma(m) \neq 0$ if and only if $v(a) + v(m) = v(am)$ if and only if $\sigma(a)\sigma(m) = \sigma(am)$ if and only if $\widetilde{a}\widetilde{m} = \widetilde{am}$.

□

3.12. Proposition Let A be a $\mathbb{Z}$-filtered ring with filtration FA.
(i) A is a domain if and only if $\widetilde{A}$ is a domain.
(ii) Suppose that FA is separated. If $G(A)$ is a domain then so are A and $\widetilde{A}$.

□

3.13. Proposition Let A be a $\mathbb{Z}$-filtered ring with filtration FA, and M a filtered A-module with a discrete filtration FM (i.e., $F_n M = 0$ for some $n \in \mathbb{Z}$). If $G(M)$ is Noetherian then so are M and $\widetilde{M}$.

□

If A is a positively filtered ring such that $G(A)$ is Noetherian, then the above proposition entails that A and $\widetilde{A}$ are Noetherian. This is a special case of the so called *Zariskian filtered ring* in the sense of ([Li1], [LVO4]). For a Zariskian filtered ring A, many algebraic structure properties of $G(A)$ may be lifted to A.

This shows on one hand the advantage of using the associated graded structures. We refer to [LVO4] for a general theory of Zariskian filtrations.

Let A be a filtered ring with filtration FA, and let M, N be filtered A-modules with filtrations FM, FN, respectively. An A-homomorphism $\varphi\colon M \to N$ is said to be a *filtered A-homomorphism* if $\varphi(F_pM) \subset F_pN$ for all $p \in \mathbb{Z}$. If φ is an A-homomorphism such that $\varphi(F_pM) = F_pN \cap \operatorname{Im}\varphi$ for all $p \in \mathbb{Z}$, we say that φ is a *strict filtered A-homomorphism*.

Let $\varphi\colon M \to N$ be a filtered A-homomorphism. Then φ induces two graded module homomorphisms in a natural way:

$$G(M) \xrightarrow{G(\varphi)} G(N), \qquad \widetilde{M} \xrightarrow{\widetilde{\varphi}} \widetilde{N}.$$

Let M be a filtered A-module with filtration FM, and N a submodule of M. Then the *induced filtration* on N is defined as: $F_pN = F_pM \cap N$, $p \in \mathbb{Z}$. The *induced quotient filtration* on M/N is defined as: $F_p(M/N) = (F_pM + N)/N$, $p \in \mathbb{Z}$.

3.14. Proposition Let $\varphi\colon M \to N$ be an onto strict filtered A-homomorphism with $\operatorname{Ker}\varphi = K$. Consider the induced filtration on K and let $\ell\colon K \to M$ be the inclusion map. Then the exact sequence of A-modules

$$0 \to K \xrightarrow{\ell} M \xrightarrow{\varphi} N \to 0$$

induces two exact sequences of graded modules:

$$0 \to G(K) \xrightarrow{G(\ell)} G(M) \xrightarrow{G(\varphi)} G(N) \to 0$$

$$0 \to \widetilde{K} \xrightarrow{\widetilde{\ell}} \widetilde{M} \xrightarrow{\widetilde{\varphi}} \widetilde{N} \to 0$$

$\square$

3.15. Corollary Let M be a filtered A-module with filtration FM, and N a submodule of M. Consider the filtrations on N and M/N induced by FM, respectively. The following holds.

(i) $G(N)$ is a graded submodule of $G(M)$, and $\widetilde{N}$ is a graded submodule of $\widetilde{M}$.

(ii) $G(M/N) \cong G(M)/G(N)$, $\widetilde{M/N} \cong \widetilde{M}/\widetilde{N}$.

$\square$

Example (iii) Let $R = \oplus_{n \in \mathbb{Z}} R_n$ be a $\mathbb{Z}$-graded ring. Then the *grading filtration* FR on R is defined as

$$F_pR = \bigoplus_{n \leq p} R_n, \quad p \in \mathbb{Z}.$$

It is easy to see that $G(R) \cong R$ as graded rings. In Example (vi) below, we will prove that $\widetilde{R} \cong R[t]$ as graded rings, where $R[t]$ has the mixed gradation (see the foregoing Example (ii)).

(iv) Let $A = k[a_i]_{i \in \Lambda}$ be a k-algebra generated by $\{a_i\}_{i \in \Lambda}$ over k. Then the naturally defined positive filtration FA on A consisting of k-subspaces

$$F_p A = \left\{ \sum_{\alpha_1 + \cdots + \alpha_n \leq p} c_\alpha a_{i_1}^{\alpha_1} \cdots a_{i_n}^{\alpha_n} \;\middle|\; c_\alpha \in k,\ \alpha_j \in \mathbb{N} \right\}, \quad p \in \mathbb{N},$$

is called the *standard filtration* on A. We note the following basic facts for the standard filtration FA.

a. If $A = k[a_1, ..., a_n]$ is a finitely generated algebra over k with the finite generating set $\{a_1, .., a_n\}$, then every $F_p A$ is a *finite* dimensional subspace of the k-vector space A. In particular, $F_0 A = k$.

If $A = k[a_i]_{i \in \Lambda} = \oplus_{n \in \mathbb{N}} A_n$ is also positively graded such that $a_i \in A_1$, $i \in \Lambda$, i.e., A is generated by homogeneous elements of degree 1, then it is easy to see that

b. $A_0 = k$, $A_1 = \sum_i k a_i$, $A_n = A_1^n$ and the standard filtration FA on A is given by the grading filtration on A as defined in (iii) above.

Concerning the generators of $G(A)$ and $\widetilde{A}$ with respect to the standard filtration on A, where $A = k[a_i]_{i \in \Lambda}$, we have the following fact.

3.16. Proposition (i) $G(A) = k[\sigma(a_i)]_{i \in \Lambda}$, i.e., $G(A)$ is generated by homogeneous elements of degree 1.
(ii) $\widetilde{A} = k[X, \widetilde{a}_i]_{i \in \Lambda}$, where X is the canonical element of $\widetilde{A}$, i.e., $\widetilde{A}$ is also generated by homogeneous elements of degree 1.

Proof (i) Since FA is the standard filtration, it is separated. If $a \in A$ is nonzero, then $a \in F_h A - F_{h-1} A$ for some $h \geq 0$, i.e., $v(a) = h$, in particular, every $\sigma(a_i)$ has degree 1.
For $a \in A$ with order $v(a) = h$, writing

$$a = \sum_{|\alpha| \leq h} c_\alpha a_{i_1}^{\alpha_1} \cdots a_{i_n}^{\alpha_n}, \quad |\alpha| = \alpha_1 + \cdots + \alpha_n,$$

$\sigma(a) = \sum_{|\alpha| = h} (c_\alpha a_{i_1}^{\alpha_1} \cdots a_{i_n}^{\alpha_n} + F_{h-1} A)$. We may assume that every coset $a_{i_1}^{\alpha_1} \cdots a_{i_n}^{\alpha_n} + F_{h-1} A \neq 0$ in the above expression is nonzero. Then by Lemma 3.11 we have

$$\sigma(a) = \sum_{|\alpha| = h} c_\alpha \sigma(a_{i_1})^{\alpha_1} \cdots \sigma(a_{i_n})^{\alpha_n}.$$

(ii) That every $\widetilde{a}_i$ is of degree 1 is clear. Let $F \in \widetilde{A}_p$ be a nonzero homogeneous element of degree p. By Lemma 3.11, $F = X^s \widetilde{a}$ for some $s \geq 0$, $a \in A$. Writing $a = \sum c_{i_1 \cdots i_n} a_{i_1}^{\alpha_1} \cdots a_{i_n}^{\alpha_n}$ and using the order function as defined before Lemma 3.11, we may   assume that $v(a) = h$ and $v(a_{i_1}^{\alpha_1} \cdots a_{i_n}^{\alpha_n}) = p_\alpha$. Put $|\alpha| = \alpha_1 +$

$\cdots + \alpha_n$. Then since FA is the standard filtration we have $|\alpha| \le h$, $p_\alpha \le h$, and by Lemma 3.11

$$X^{|\alpha|-p_\alpha}\widetilde{(a_{i_1}^{\alpha_1} \cdots a_{i_n}^{\alpha_n})} = \widetilde{a_{i_1}}^{\alpha_1} \cdots \widetilde{a_{i_n}}^{\alpha_n}.$$

Thus

$$
\begin{aligned}
\widetilde{a} &= \sum c_{i_1 \cdots i_n} X^{h-p_\alpha} \widetilde{(a_{i_1}^{\alpha_1} \cdots a_{i_n}^{\alpha_n})} \\
&= \sum c_{i_1 \cdots i_n} X^{h-|\alpha|+|\alpha|-p_\alpha} \widetilde{(a_{i_1}^{\alpha_1} \cdots a_{i_n}^{\alpha_n})} \\
&= \sum c_{i_1 \cdots i_n} X^{h-|\alpha|} \widetilde{a_{i_1}}^{\alpha_1} \cdots \widetilde{a_{i_n}}^{\alpha_n}
\end{aligned}
$$

and consequently

$$F = X^s \widetilde{a} = \sum c_{i_1 \cdots i_n} X^{h-|\alpha|+s} \widetilde{a_{i_1}}^{\alpha_1} \cdots \widetilde{a_{i_n}}^{\alpha_n}.$$

This finishes the proof. $\square$

Now let $k\langle X \rangle$ be the free k-algebra generated by $X = \{X_i\}_{i \in \Lambda}$. Then from the foregoing Example (i) we know that $k\langle X \rangle$ is a positively graded ring, i.e., $k\langle X \rangle = \oplus_{p \in I\!\!N} k\langle X \rangle_p$, and $X_i \in k\langle X \rangle_1$, $i \in \Lambda$. Hence, the standard filtration $F k\langle X \rangle$ on $k\langle X \rangle$ is given by the grading filtration on $k\langle X \rangle$: $F_p k\langle X \rangle = \oplus_{q \le p} k\langle X \rangle_q$, $p \in I\!\!N$. If we consider the natural onto algebra homomorphism $\varphi \colon k\langle X \rangle \to A = k[a_i]_{i \in \Lambda}$ with $\varphi(X_i) = a_i$, $i \in \Lambda$, then it is also clear that $\varphi(F_p k\langle X \rangle) = F_p A$, $p \in I\!\!N$, i.e., φ is a strict filtered $k\langle X \rangle$-homomorphism. If furthermore we consider the filtration on $I = \mathrm{Ker}\varphi$ induced by $F k\langle X \rangle$, then by Proposition 3.11, the exact sequence of $k\langle X \rangle$-modules

$$0 \to I \xrightarrow{\ell} k\langle X \rangle \xrightarrow{\varphi} A \to 0$$

induces two exact sequences of graded modules:

$$0 \to G(I) \xrightarrow{G(\ell)} G(k\langle X \rangle) \xrightarrow{G(\varphi)} G(A) \to 0$$

$$0 \to \widetilde{I} \xrightarrow{\widetilde{\ell}} \widetilde{k\langle X \rangle} \xrightarrow{\widetilde{\varphi}} \widetilde{A} \to 0$$

From the above Example (iii) we know that $G(k\langle X \rangle) \cong k\langle X \rangle$. In Example (vi) below it will be shown that $\widetilde{k\langle X \rangle} \cong k\langle X \rangle[t]$ as graded rings, where $k\langle X \rangle[t]$ is the polynomial ring over $k\langle X \rangle$ in one commuting variable t, and the gradation used on $k\langle X \rangle[t]$ is the mixed gradation as defined in the foregoing Example (ii). Later in CH.III we will see that if the the defining relations of A are given, namely, a generating set of the above I is given, then it is possible to algorithmically determine the generating sets for both $G(I)$ and $\widetilde{I}$.

Example (v) This example may be viewed as the dual process of constructing the Rees ring of a filtered ring with a given filtration. The first general study of this process and some applications were given in [LVO3].

Let $R = \oplus_{n \in \mathbb{Z}} R_n$ be a $\mathbb{Z}$-graded ring. Suppose that R contains a homogeneous element Y of degree 1, i.e., $Y \in R_1$, which is contained in the *centre* of R and is *not* a divisor of zero. Then it is clear that both $\langle Y \rangle = YR$ and $\langle 1-Y \rangle = (1-Y)R$ are two-sided ideals in R, in particular, $\langle Y \rangle$ is a graded ideal of R by Proposition 3.3.
Put

$$A = R/\langle 1 - Y \rangle.$$

Note that if $r_n \in R_n$ is a homogeneous element of degree n, then $Yr_n \in R_{n+1}$ because Y is of degree 1. Thus $r_n = Yr_n + (1-Y)r_n$ implies $R_n + \langle 1-Y \rangle \subset R_{n+1} + \langle 1-Y \rangle$. Consequently we have

$$\frac{R_n + \langle 1-Y \rangle}{\langle 1-Y \rangle} \subset \frac{R_{n+1} + \langle 1-Y \rangle}{\langle 1-Y \rangle}, \quad n \in \mathbb{Z},$$

$$\bigcup_{n \in \mathbb{Z}} \frac{R_n + \langle 1-Y \rangle}{\langle 1-Y \rangle} = \frac{R}{\langle 1-Y \rangle} = A.$$

Hence the $\mathbb{Z}$-gradation on R naturally induces a $\mathbb{Z}$-filtration FA on A, i.e.,

$$F_n A = \frac{R_n + \langle 1-Y \rangle}{\langle 1-Y \rangle}, \quad n \in \mathbb{Z}.$$

3.17. Proposition With notation as above, the following holds.
(i) $\langle 1-Y \rangle$ does not contain any nonzero homogeneous element of R.
(ii) $G(A) \cong R/\langle Y \rangle$ as graded rings.
(iii) $\tilde{A} \cong R$ as graded rings, where Y corresponds to the canonical element X in $\tilde{A}$ (see the definition given before Proposition 3.8).

Proof (i) Since Y is a homogeneous element of degree 1 in the center of R and it is not a divisor of zero, this may be verified directly.
(ii) By definition

$$G(A)_n \cong \frac{R_n + \langle 1-Y \rangle}{R_{n-1} + \langle 1-Y \rangle}, \quad n \in \mathbb{Z}.$$

Note that since $R/\langle Y \rangle = \oplus_{n \in \mathbb{Z}} (R/\langle Y \rangle)_n$ is a graded ring (see Proposition 3.3) with

$$(R/\langle Y \rangle)_n = \frac{R_n + \langle Y \rangle}{\langle Y \rangle}, \quad n \in \mathbb{Z},$$

If for each $n \in \mathbb{Z}$ we define the map

$$\varphi_n : \quad \frac{R_n + \langle Y \rangle}{\langle Y \rangle} \quad \longrightarrow \quad \frac{R_n + \langle 1-Y \rangle}{R_{n-1} + \langle 1-Y \rangle}$$

$$r_n + \langle Y \rangle \quad \mapsto \quad r_n + (R_{n-1} + \langle 1-Y \rangle)$$

then it is not hard to see that these maps yield a graded ring isomorphism $\varphi = \oplus \varphi_n \colon R/\langle Y \rangle \to G(A)$.

(iii) By definition

$$\tilde{A} = \oplus_{n \in \mathbb{Z}} F_n A = \bigoplus_{n \in \mathbb{Z}} \frac{R_n + \langle 1 - Y \rangle}{\langle 1 - Y \rangle}.$$

It follows from (i) above that for each $n \in \mathbb{Z}$ there is a natural group isomorphism:

$$\psi_n \colon \quad R_n \quad \longrightarrow \quad \frac{R_n + \langle 1 - Y \rangle}{\langle 1 - Y \rangle}$$

$$r_n \quad \longmapsto \quad r_n + \langle 1 - Y \rangle$$

It is easy to verify that these ψ_n yield a graded ring isomorphism $\psi = \oplus \psi_n \colon R \to \tilde{A}$, in particular, $\psi(Y) = X$ where the latter is the canonical element of $\tilde{A}$. $\qquad \square$

Example (vi) Let $R = \oplus_{n \in \mathbb{Z}} R_n$ be a $\mathbb{Z}$-graded ring, and consider the grading filtration FR on R as defined in Example (iii) above, i.e., $F_n R = \oplus_{i \leq n} R_i$, $n \in \mathbb{Z}$. We have seen that $G(R) \cong R$. The aim of this example is to determine $\tilde{R}$.

Let $R[t]$ be the polynomial ring over R in one commuting variable t, and consider the mixed gradation on $R[t]$ as defined in the foregoing Example (ii), i.e., $R[t] = \oplus_{n \in \mathbb{Z}} R[t]_n$ with

$$R[t]_n = \left\{ \sum_{i+j=n} a_i t^j \ \middle| \ a_i \in R_i, \ j \in \mathbb{N} \right\}, \quad n \in \mathbb{Z}.$$

Note that t is a homogeneous element of degree 1 contained in the centre of $R[t]$ and it is not a divisor of zero. It follows from Example (v) above that the mixed gradation on $R[t]$ induces a filtration FA on $A = R[t]/\langle 1 - t \rangle$, i.e.,

$$F_n A = \frac{R[t]_n + \langle 1 - t \rangle}{\langle 1 - t \rangle}, \quad n \in \mathbb{Z}.$$

Now we have the following easily verified facts:

c. The onto ring homomorphism $\phi \colon R[t] \to R$ with $\phi(t) = 1$ has $\mathrm{Ker}\phi = \langle 1 - t \rangle$, hence $A = R[t]/\langle 1 - t \rangle \cong R$ as rings,

d. $F_n A = \dfrac{R[t]_n + \langle 1 - t \rangle}{\langle 1 - t \rangle} \cong R[t]_n$ as additive groups by Proposition 3.17(i), and

e. $\oplus_{i \leq n} R_i = F_n R \cong F_n A = \dfrac{R[t]_n + \langle 1 - t \rangle}{\langle 1 - t \rangle}$ as additive groups, where each $\sum_{i \leq n} a_i \in F_n R$ is sent to $\sum a_i t^{n-i} + \langle 1 - t \rangle$.

Hence, it follows from Proposition 3.17(ii)–(iii) that the following holds.

3.18. Proposition Let $R = \oplus_{n \in \mathbb{Z}} R_n$ be a $\mathbb{Z}$-graded ring and FR the grading filtration on R. Then
(i) $G(R) \cong R$ as graded rings,
(ii) $\tilde{R} \cong \tilde{A} \cong R[t]$ as graded rings.

$\square$

4. Homogenization and Dehomogenization of $\mathbb{Z}$-graded Rings

Let k be an algebraically closed field of char$k = 0$, and $k[x_1, ..., x_n]$ the commutative polynomial k-algebra in n variables. If I is an ideal of $k[x_1, ..., x_n]$ and I^* is the homogenization ideal of I in $k[x_0, x_1, ..., x_n]$ with respect to x_0, then a well known fact from algebraic geometry is that I^* defines the projective closure of the affine algebraic set $V(I)$ in the projective n-space $\mathbf{P}_k^n$. In this section we will see that there is a nice relation between the algebra $k[x_1, ..., x_n]/I$ and the algebra $k[x_0, x_1, ..., x_n]/I^*$, which has a *very noncommutative* version, and the latter indeed yields more elegant structural properties describing the associated graded structures of a $\mathbb{Z}$-filtered ring.

Let $R = \oplus_{n \in \mathbb{Z}} R_n$ be a $\mathbb{Z}$-graded ring and $R[t]$ the polynomial ring over R in one commuting variable t. Consider the familiar onto ring homomorphism

$$\phi : \quad R[t] \longrightarrow R$$

with $\phi(t) = 1$. We know that $\mathrm{Ker}\phi = \langle 1 - t \rangle$ and hence $R \cong R[t]/\langle 1 - t \rangle$. Considering the mixed gradation on $R[t]$ as defined in §4 Example (ii), i.e., $R[t] = \oplus_{p \in \mathbb{Z}} R[t]_p$ with

$$R[t]_p = \left\{ \sum_{i+j=p} F_i t^j \ \middle| \ F_i \in R_i, \ j \geq 0 \right\}, \quad p \in \mathbb{Z},$$

we also observe that

- For every $f \in R$, there exists a homogeneous element $F \in R[t]_p$, for some p, such that $\phi(F) = f$. More precisely, if $f = F_0 + F_1 + \cdots + F_p$ where $F_i \in R_i$, $F_p \neq 0$, then $f^* = t^p F_0 + t^{p-1} F_1 + \cdots + t F_{p-1} + F_p$ is a homogeneous element in $R[t]_p$ satisfying $\phi(f^*) = f$.

4.1. Definition (i) For any $F \in R[t]$, we write $F_* = \phi(F)$. F_* is called the *dehomogenization* of F with respect to t.

(ii) For any $f \in R$, if $f = F_0 + F_1 + \cdots + F_p$ with $F_p \neq 0$, then the homogeneous element $f^* = t^p F_0 + t^{p-1} F_1 + \cdots + t F_{p-1} + F_p$ in $R[t]_p$ is called the *homogenization* of f with respect to t.

(iii) If I is a two-sided ideal of R, then we let I^* stand for the graded two-sided ideal of $R[t]$ generated by $\{f^* \mid f \in I\}$. I^* is called the *homogenization ideal* of I with respect to t.

For $f \in R$, if $f = F_0 + F_1 + \cdots + F_p$ is the homogeneous decomposition of f and $F_p \neq 0$, we write $d(f) = p$.

Note that since t is a commuting variable, the following lemma can be easily verified.

4.2. Lemma (i) For $F, G \in R[t]$, $(F + G)_* = F_* + G_*$, $(FG)_* = F_* G_*$.

(ii) For $f, g \in R$, $(fg)^* = f^* g^*$, $t^s (f + g)^* = t^r f^* + t^h g^*$, where $r = d(g)$, $h = d(f)$, and $s = r + h - d(f + g)$.

(iii) For any $f \in R$, $(f^*)_* = f$.

(iv) If F is a homogeneous element of degree p in $R[t]$, and if $(F_*)^*$ is of degree q, then $t^r (F_*)^* = F$, where $r = p - q$.

(v) If I is a two-sided ideal of R, then each homogeneous element $F \in I^*$ is of the form $t^r f^*$ for some $r \in \mathbb{N}$ and $f \in I$.

$\square$

4.3. Proposition Let I be a proper two-sided ideal of R. Then there is an onto ring homomorphism $\alpha \colon R[t]/I^* \to R/I$ with $\mathrm{Ker}\,\alpha = \langle 1 - \bar{t} \rangle$, where $\bar{t}$ denotes the coset of t in $R[t]/I^*$. Moreover, $\bar{t}$ is not a divisor of zero in $R[t]/I^*$, and hence $\langle 1 - \bar{t} \rangle$ does not contain any nonzero homogeneous element of $R[t]/I^*$.

Proof If we define α by putting

$$\alpha \colon \quad \begin{array}{ccc} R[t]/I^* & \longrightarrow & R/I \\ F + I^* & \mapsto & F_* + I, \end{array} \quad F \in R[t],$$

then by Lemma 4.2 we easily see that α is a ring epimorphism with $\mathrm{Ker}\,\alpha = \langle 1 - \bar{t} \rangle$. For any homogeneous element $F \in R[t]$, if $tF \in I^*$, then $F_* = (tF)_* \in (I^*)_* \subset I$ by Lemma 4.2. Again by Lemma 4.2 we have $F = t^r (F_*)^* \in I^*$. Hence $\bar{t}$ is not a divisor of zero in $R[t]/I^*$. The fact that $\langle 1 - \bar{t} \rangle$ does not contain any nonzero homogeneous element of $R[t]/I^*$ now follows easily. $\square$

Now let $k\langle X \rangle$ be the free algebra over a field k with generating set $X = \{X_i\}_{i \in \Lambda}$. From §4 we retain that $k\langle X \rangle = \oplus_{p \in \mathbb{N}} k\langle X \rangle_p$ has the natural gradation defined by the degree of words in the free semigroup $S = \langle X \rangle$ and the standard filtration $F k\langle X \rangle$ on $k\langle X \rangle$ is given by the k-subspaces:

$$F_p k\langle X \rangle = \oplus_{i \leq p} k\langle X \rangle_i, \quad p \in \mathbb{N}.$$

If I is any two-sided ideal of $k\langle X\rangle$ and we put $A = k\langle X\rangle/I$, then $Fk\langle X\rangle$ induces the quotient filtration FA on A:

$$F_pA = (F_pk\langle X\rangle + I)/I, \quad p \in I\!N,$$

which indeed coincides with the standard filtration on the k-algebra $A = k\langle X\rangle/I = k[x_i]_{i\in\Lambda}$ where x_i is the coset of X_i in $k\langle X\rangle/I$.

Let $k\langle X\rangle[t]$ be the polynomial ring over $k\langle X\rangle$ in one commuting variable t and consider the mixed gradation on $k\langle X\rangle[t]$. If we consider the associated graded ring $G(A) = \oplus_{p\geq 0}(F_pA/F_{p-1}A)$ of A and the Rees ring $\widetilde{A} = \oplus_{p\geq 0}F_pA$ of A, then the following proposition shows that $G(A)$ and $\widetilde{A}$ are completely determined by I^*, where I^* is the homogenization ideal of I in $k\langle X\rangle[t]$ (in the sense of Definition 4.1).

4.4. Proposition (compare with Proposition 3.16) With notation as above, there are graded k-algebra isomorphisms:
(i) $\widetilde{A} \cong k\langle X\rangle[t]/I^*$, and
(ii) $G(A) \cong k\langle X\rangle[t]/(\langle t\rangle + I^*)$, where $\langle t\rangle$ denotes the ideal of $k\langle X\rangle[t]$ generated by t.

Proof First note that the ring homomorphism α: $k\langle X\rangle[t]/I^* \rightarrow k\langle X\rangle/I = A$ defined in Proposition 4.3 yields the isomorphism of k-subspaces:

$$\frac{k\langle X\rangle[t]_p + \langle 1 - \bar{t}\rangle}{\langle 1 - \bar{t}\rangle} \longrightarrow \frac{\oplus_{i\leq p}k\langle X\rangle_i + I}{I} = F_pA, \quad p \in I\!N.$$

Since we know from Proposition 4.3 that $\bar{t}$ is a homogeneous element of degree 1 in $k\langle X\rangle[t]/I^*$ and it is not a divisor of zero, it follows from Proposition 3.8 that (i) and (ii) hold. □

Remark Let us go back to the commutative case and consider $A = k[x_1, ..., x_n]/I$, where I is an ideal of the polynomial algebra $k[x_1, ..., x_n]$. Now it is clear that, with respect to the standard filtration FA on A (which is in fact induced by the natural grading filtration on $k[x_1, ..., x_n]$), $G(A) \cong k[x_0, x_1, ..., x_n]/(\langle x_0\rangle + I^*)$ and $\widetilde{A} \cong k[x_0, x_1, ..., x_n]/I^*$, where I^* is the homogenization ideal of I in $k[x_0, x_1, ..., x_n]$ with respect to x_0. This implies that the defining relations of the Rees algebra $\widetilde{A}$ of A correspond to the defining equations of the projective closure $V(I^*)$ of the affine algebraic set $V(I)$, and the defining relations of the associated graded ring $G(A)$ correspond to the defining equations of the part of the projective closure $V(I^*)$ at infinity.

5. Some Algebras: Classical and Modern

In this section, we list some well-known algebras as the basic examples in developing later chapters. More examples are constructed in CH.III.

(i) Weyl algebra

The first Weyl algebra $A_1(k)$ over a field k is defined to be the k-algebra generated by x and y subject to the relation

$$yx = xy + 1.$$

Let $k[t]$ be the polynomial algebra over k in t and $\delta = \partial/\partial t$. It follows from §2 Example (i) that

$$A_1(k) \cong k[t][z; \delta].$$

The nth Weyl algebra $A_n(k)$ over a field k is defined to be the k-algebra generated by $2n$ generators $x_1, ..., x_n, y_1, ..., y_n$ subject to the relations:

$$x_i x_j = x_j x_i, \ y_i y_j = y_j y_i, \qquad\qquad 1 \le i < j \le n,$$
$$y_j x_i = x_i y_j + \delta_{ij} \text{ the Kronecker delta}, \ \ 1 \le i, j \le n.$$

If we write $R = k[t_1, ..., t_n]$ for the commutative polynomial algebra in n variables, δ_i for $\partial/\partial t_i$, and consider the sequence of skew polynomial algebras:

$$R_0 = R, \quad R_{i+1} = R_i[z_{i+1}; \delta_{i+1}], \ i = 0, ..., n-1,$$

then the k-algebra R_n has generators which satisfy the relations that define the Weyl algebra (note that since for each i,

$$\left\{ t_1^{\alpha_1} \cdots t_n^{\alpha_n} z_1^{\beta_1} \cdots z_i^{\beta_i} \ \middle| \ (\alpha_1, ..., \alpha_n, \beta_1, ..., \beta_i) \in I\!N^{n+i} \right\}$$

is a k-basis of R_i and t_{i+1} is in the centre of R_i, it is easy to see that $\partial/\partial t_{i+1}$ defines a derivation on R_i). On the other hand, the generators of $A_n(k)$ satisfy the defining relations for the R_i. Thus, one easily construct an algebra isomorphism $A_n(k) \cong R_n$, i.e., $A_n(k)$ is indeed an iterated skew polynomial ring (or an iterated Ore extension) on $k[t_1, ..., t_n]$.

Historically, the Weyl algebra is the first "quantum algebra" ([Dir] 1926, [Wey] 1928). In the literature the standard filtration (see §3) on $A_n(k)$ is customarily called the Bernstein filtration because of the celebrated paper of I.N Bernstein [Bern]. It is well-known that, with respect to the standard filtration on $A_n(k)$, the associated graded ring $G(A_n(k))$ of $A_n(k)$ is isomorphic to the commutative polynomial ring over k in $2n$ variables (see, e.g., [Bj]). In CH.III §3 we will recapture this structure in an algorithmic way, and we will also give an algorithmic determination of the defining relations of the associated Rees algebra $\widetilde{A_n}(k)$.

Another well-known fact about $A_n(k)$ is that, if $\operatorname{char} k = 0$, then $A_n(k)$ coincides with the algebra of linear partial differential operators of the polynomial   ring

$k[t_x, ..., t_n]$ (or the ring of polynomial functions of the affine n-space $\mathbf{A}_k^n$). We will come back to this point in CH.II §8 and CH.VII.

(ii) Additive analogue of the Weyl algebra

This algebra was introduced in Quantum Physics in ([Kur] 1980) and studied in ([JBS] 1981), that is, the algebra $A_n(q_1, ..., q_n)$ generated over a field k by $x_1, ..., x_n, y_1, ..., y_n$ subject to the relations:

$$
\begin{aligned}
x_i x_j = x_j x_i, \ y_i y_j = y_j y_i, & \quad 1 \le i < j \le n, \\
y_i x_i = q_i x_i y_i + 1, & \quad 1 \le i \le n, \\
x_j y_i = y_i x_j, & \quad i \ne j,
\end{aligned}
$$

where $q_i \in k - \{0\}$. It is not hard to see that $A_n(q_1, ..., q_n)$ is isomorphic to the iterated skew polynomial ring on the commutative polynomial ring $k[t_1, ..., t_n]$:

$$k[t_1, ..., t_n][z_1; \sigma_1, \delta_1] \cdots [z_n; \sigma_n, \delta_n],$$
where
$$
\begin{aligned}
z_j z_i = z_i z_j, & \quad 1 \le i < j \le n, \\
z_i t_i = \sigma_i(t_i) z_i + \delta_i(t_i) = q_i t_i z_i + 1, & \quad 1 \le i \le n, \\
z_i t_j = t_j z_i, & \quad i \ne j.
\end{aligned}
$$

If $q_i = q \ne 0$, $i = 1, ..., n$, then this algebra becomes the algebra of q-*differential operators*. In CH.VII we will come back to this point.

(iii) Multiplicative analogue of the Weyl algebra

This is the algebra stemming from ([Jat] 1984 and [MP] 1988, where one may see why this algebra deserves its title), that is, the algebra $\mathcal{O}_n(\lambda_{ji})$ generated over a field k by $x_1, ..., x_n$ subject to the relations:

$$x_j x_i = \lambda_{ji} x_i x_j, \quad 1 \le i < j \le n,$$

where $\lambda_{ji} \in k - \{0\}$. It is easy to see that $\mathcal{O}_n(\lambda_{ji})$ is isomorphic to the iterated skew polynomial ring

$$k[z_1][z_2; \sigma_2] \cdots [z_n; \sigma_n],$$
where $z_j z_i = \sigma_j(z_i) z_j = \lambda_{ji} z_i z_j, \ 1 \le i < j \le n.$

If $n = 2$, then $\mathcal{O}_2(\lambda_{21})$ is the *quantum plane* in the sense of Manin [Man]. If $\lambda_{ji} = q^{-2} \ne 0$ for some $q \in k - \{0\}$ and all $1 \le i < j \le n$, then $\mathcal{O}_n(\lambda_{ji})$ becomes the well-known coordinate ring of the so called *quantum affine n-space* (e.g., see [Sm1]).

Algebras of type $\mathcal{O}_n(\lambda_{ji})$ will play a key role in the structural-algorithmic study of quadric solvable polynomial algebras (CH.III–CH.VIII), and for the sake of using a unified notion, these algebras will be called homogeneous solvable polynomial algebras (CH.III definition 2.1).

(iv) Enveloping algebra of a finite dimensional Lie algebra

Let $\mathbf{g}$ be a finite dimensional vector space over the field k with basis $\{x_1, ..., x_n\}$ where $n = \dim_k \mathbf{g}$. If there is a binary operation on $\mathbf{g}$, called the *bracket product* and denoted $[\ ,\ ]$, which is bilinear, i.e., for $a, b, c \in \mathbf{g}$, $\lambda \in k$,

$$[a + b, c] = [a, c] + [b, c]$$
$$[a, b + c] = [a, b] + [a, c]$$
$$\lambda[a, b] = [\lambda a, b] = [a, \lambda b],$$

and satisfies:

$$[a, b] = -[b, a], \quad a, b \in \mathbf{g},$$
$$[[a, b], c] + [[c, a], b] + [[b, c], a] = 0 \quad \text{Jacobi identity, } a, b, c \in \mathbf{g},$$

then $\mathbf{g}$ is called a finite dimensional Lie algebra over k. Note that $[\ ,\ ]$ need not satify the associative low. If $[a, b] = [b, a]$ for every a, b in a Lie algebra $\mathbf{g}$, $\mathbf{g}$ is called *abelian*.

The enveloping algebra of $\mathbf{g}$, denoted $U(\mathbf{g})$, is defined to be the *associative k-algebra* generated by $x_1, ..., x_n$ subject to the relations:

$$x_j x_i - x_i x_j = [x_j, x_i], \quad 1 \leq i < j \leq n.$$

For example, the Heisenberg Lie algebra $\mathbf{h}$ has the k-basis $\{x_i, y_j, z \mid i, j = 1, ..., n\}$ and the bracket product is given by

$$[x_i, y_i] = z, \qquad\qquad\qquad 1 \leq i \leq n,$$
$$[x_i, x_j] = [x_i, y_j] = [y_i, y_j] = 0, \quad i \neq j,$$
$$[z, x_i] = [z, y_i] = 0, \qquad\qquad 1 \leq i \leq n.$$

In Example (v) below, one will see that the enveloping algebra $U(\mathbf{h})$ of the Heisenberg Lie algebra $\mathbf{h}$ is an iterated skew polynomial ring.

If we consider the standard filtration on $U(\mathbf{g})$, then it follows from the famous Poincaré-Birkhoff-Witt theorem (abbreviated PBW theorem hereafter) that $G(U(\mathbf{g}))$ is (as a graded ring) isomorphic to the commutative polynomial ring $k[t_1, ..., t_n]$ (see, e.g., [Dix]). In CH.III §3 we will recapture this structure in an algorithmic way, and we will also give an algorithmic determination of the defining relations of the associated Rees algebra $\widetilde{U(\mathbf{g})}$.

(v) q-Heisenberg algebra

This is the algebra stemming from ([Ber] 1992, [Ros] 1995) which has its root in q-calculus (e.g., [Wal] 1985), that is, the algebra $\mathbf{h}_n(q)$ generated over a field k by the set of elements $\{x_i, y_i, z_i \mid i = 1, ..., n\}$ subject to the relations:

$$x_i x_j = x_j x_i, \ y_i y_j = y_j y_i, \ z_j z_i = z_i z_j, \quad 1 \leq i < j \leq n,$$
$$x_i z_i = q z_i x_i, \qquad\qquad\qquad\qquad 1 \leq i \leq n,$$
$$z_i y_i = q y_i z_i, \qquad\qquad\qquad\qquad 1 \leq i \leq n,$$
$$x_i y_i = q^{-1} y_i x_i + z_i, \qquad\qquad\quad 1 \leq i \leq n,$$
$$x_i y_j = y_j x_i, \ x_i z_j = z_j x_i, \ y_i z_j = z_j y_i, \quad i \neq j,$$

where $q \in k - \{0\}$. It is not hard to see that $\mathbf{h}_n(q)$ is isomorphic to the iterated skew polynomial ring on the commutative polynomial ring $k[t_1, ..., t_n]$:

$$k[t_1, ..., t_n][u_1; \sigma_1] \cdots [u_n; \sigma_n][v_1; \theta_1, \delta_1] \cdots [v_n; \theta_n, \delta_n],$$

where

$$
\begin{array}{ll}
v_i v_j = v_j v_i, \quad u_i u_j = u_j u_i, & i \neq j, \\
u_i t_i = \sigma(t_i) u_i = q t_i u_i, & 1 \leq i \leq n, \\
v_i u_i = \theta(u_i) v_i = q u_i v_i, & 1 \leq i \leq n, \\
v_i t_i = \theta(t_i) v_i + \delta_i(t_i) = q^{-1} t_i v_i + u_i, & 1 \leq i \leq n, \\
u_j t_i = t_i u_j, \quad v_j t_i = t_i v_j, \quad v_j u_i - u_i v_j, & i \neq j.
\end{array}
$$

(vi) Manin algebra of 2×2 quantum matrices

See ([Man] 1988). This is the algebra $M_q(2, k)$ generated over a field k by a, b, c, d subject to the relations:

$$
\begin{array}{lll}
ba = qab, & ca = qac, & dc = qcd, \\
db = qbd, & cb = bc, & da - ad = (q - q^{-1})bc,
\end{array}
$$

where $q \in k - \{0\}$. Note that this is a positively graded quadratic algebra with the natural gradation (see §3), and it is is also an iterated skew polynomial algebra starting with the ground field k (see CH.III Example (vi))

(vii) Hayashi algebra

In order to get bosonic representations for the types of $\mathbf{A}_n$ and $\mathbf{C}_n$ of the Drinfield-Jimbo quantum algebras, Hayashi introduced in ([Hay] 1990) the q-Weyl algebra $\mathcal{A}_q^-$, which is constructed as follows, by following [Ber]. Let $\mathbf{U}$ be the algebra generated over the field $k = \mathbb{C}$ by the set of elements $\{x_i, y_i, z_i \mid i = 1, ..., n\}$ subject to the relations:

$$
\begin{array}{ll}
x_j x_i = x_i x_j, \quad y_j y_i = y_i y_j, \quad z_j z_i = z_i z_j, & 1 \leq i < j \leq n, \\
x_j y_i = q^{-\delta_{ij}} y_i x_j, \quad z_j x_i = q^{-\delta_{ij}} x_i z_j, & 1 \leq i, j \leq n, \\
z_j y_i = y_i z_j, & i \neq j, \\
z_i y_i - q^2 y_i z_i = -q^2 x_i^2, & 1 \leq i \leq n,
\end{array}
$$

where $q \in k - \{0\}$. Then $\mathcal{A}_q^- = S^{-1}\mathbf{U}$, the localization of $\mathbf{U}$ at the multiplicative monoid S generated by $x_1, ..., x_n$. Note that $\mathbf{U}$ is a positively graded quadratic algebra with the natural gradation (see §3), and it is also an iterated skew polynomial algebra starting with the ground field k (see CH.III Example (vi)).

(viii) Grassmann Algebra

The Grassmann algebra (or exterior algebra, see, e.g., [Vdw]) is the algebra $G(\mathbf{x})$ generated over a field k by the set of elements $\mathbf{x} = \{x_i\}_{i \in \Lambda}$ subject to the relations:

$$
\begin{array}{ll}
x_i^2 = 0, & i \in \Lambda, \\
x_j x_i = -x_i x_j, & j > i,
\end{array}
$$

where a well-ordering is assumed on Λ. Obviously, $G(\mathbf{x})$ has nontrivial divisors of zero; and moreover, if $|\mathbf{x}| = n$ is finite, then $G(\mathbf{x})$ is isomorphic to $\mathcal{O}_n(\lambda_{ji})/\langle x_i^2 \rangle_{i=1}^n$, where $\mathcal{O}_n(\lambda_{ji})$ is the algebra given in Example (iii) above but with all $\lambda_{ji} = -1$, a root of unity. (A proof of the latter assertion follows from Example (ii) of CH.II §5, also see the proof given in CH.II §7 Example (i).)

(ix) Down-up Algebra

Based on the study of algebras generated by the down and up operators on a differential or uniform partially ordered set (poset), Benkart and Roby introduced in ([Ben], [BR], 1998) the down-up algebra $A(\alpha, \beta, \gamma)$, which is the algebra generated by $\{u, d\}$ over a field k subject to the relations:

$$
\begin{aligned}
d^2 u &= \alpha dud + \beta u d^2 + \gamma d \\
du^2 &= \alpha udu + \beta u^2 d + \gamma u,
\end{aligned}
$$

where $\alpha, \beta, \gamma \in k$.

CHAPTER II
Gröbner Bases in
Associative Algebras

Throughout the present chapter–CH.VIII, *most notation and notions adopt those commonly used in the quoted literature concerning Gröbner bases and related computation*, so that the reader may refer to similar texts more conveniently. In particular, to better understand this chapter, we refer the reader to [AdL], [BW], and [CLO'] for a commutative Gröbner basis theory, and to [Gr], [K-RW], and [Mor1-2] for typical noncommutative Gröbner basis theories.

The first three sections of this chapter introduces Gröbner bases for (both *two-sided* and *left*) ideals in a (not necessarily commutative) associative k-algebra $A = k[a_i]_{i \in \Lambda}$ over the fixed field k of *arbitrary* characteristic. This is reached by developing a division algorithm (§2) that is effective for any (left) admissible system $(A, \mathcal{B}, \preceq)$ including the case where A has divisors of zero. The technical path is to have a suitable "monomial ordering", distinguish the "quasi-zero" elements in A, and implement the division by using the (left) "division leading monomial" of an element $f \in A$.

In §4 we indicate the main differences of (left) Gröbner bases in various contexts, and in §5 we discuss the possibility of having a version of Buchberger's algorithm for a given (left) admissible system. With the theory developed in §§2–3, we mimic the commutative case using Dickson basis ([BW] Ch.5) to characterize the (left) admissible system $(A, \mathcal{B}, \preceq)$ in which finite (left) Gröbner bases always exist for (left) ideals of A. From an algebraic structure point of view, such algebras A will be called ("left G-Noetherian") "G-Noetherian" algebras; from an algorithmic point of view (though we are not really dealing with algorithms), we name such (left) admissible systems as (left) "Dickson systems" (§6). §7 is for presenting the class of solvable polynomial algebras (in the sense of Kandri-Rody and Weispfenning) that holds a successful Gröbner basis theory in the noncommutative setting. The chapter is closed by a closer look at the algebra of linear partial differential operators with polynomial coefficients over a field k with char$k = p > 0$, on which a Gröbner basis theory (in the sense of §3) does not exist.

1. (Left) Monomial Orderings and (Left) Admissible Systems

Let k be a field and $A = k[a_i]_{i \in \Lambda}$ a k-algebra with generating set $\{a_i\}_{i \in \Lambda}$. Then every element $f \in A$ is of the form

$$f = \sum c_{i_1 \cdots i_n} a_{i_1}^{\alpha_1} \cdots a_{i_n}^{\alpha_n}, \quad c_{i_1 \cdots i_n} \in k, \ a_{i_j} \in \{a_i\}_{i \in \Lambda}, \ \alpha_j \in I\!N.$$

Learning how to "input" and "output" polynomials of $k[t]$ in implementing the familiar Euclidean division algorithm on computer, we see that the first step to have a feasible division algorithm for A, in which the data of each "input" and "output" is ordered "compatibly" with the multiplication of A, is to define a suitable "monomial ordering" $\preceq$ on a k-basis of A. That is the goal of this section.

By abusing language, elements of the form $a_{i_1}^{\alpha_1} \cdots a_{i_n}^{\alpha_n}$ in A are called *monomials*. Let $\mathcal{B}$ be a fixed k-basis of A consisting of 1 and monomials, that is,

$$\mathcal{B} = \left\{ a_{i_1}^{\alpha_1} \cdots a_{i_n}^{\alpha_n} \ \middle| \ n \geq 1, \ \alpha_i \in I\!N \right\}.$$

To be convenient, from now on we use characters $s, \ t, \ u, \ v, \ w, \ldots$ to denote the monomials in $\mathcal{B}$.
If $w \in \mathcal{B}$, $w = a_{i_1}^{\alpha_1} \cdots a_{i_n}^{\alpha_n}$, then we write

$$|\alpha| = \alpha_1 + \cdots + \alpha_n,$$

and call $|\alpha|$ the *degree* of w, denoted $d(w) = |\alpha|$.

Let $\preceq$ denote a fixed *well-ordering* on $\mathcal{B}$ (of course, any set can be well-ordered). Then any nonzero element $f \in A$ has a *unique ordered* linear expression in terms of $w_i \in \mathcal{B}$:

$$f = \sum_{i=1}^{n} c_i w_i, \text{ with } c_i \in k - \{0\}, \text{ and } w_1 \succ w_2 \succ \cdots \succ w_n.$$

1.1. Definition Let f be a nonzero element in A as above.
(i) The *degree* of f is defined as

$$d(f) = \max\Big\{ d(w_i) \;\Big|\; i = 1, .., n \Big\}.$$

(ii) With respect to $\preceq$ we write

$$\mathbf{LM}_{\preceq}(f) = w_1, \text{ for the } \textit{leading monomial} \text{ of } f,$$
$$\mathbf{LC}_{\preceq}(f) = c_1, \text{ for the } \textit{leading coefficient} \text{ of } f,$$
$$\mathbf{LT}_{\preceq}(f) = c_1 w_1, \text{ for the } \textit{leading term} \text{ of } f.$$

If there is no confusion possible for the monomial ordering be of used, we will also just use $\mathbf{LM}(f)$, $\mathbf{LC}(f)$, and $\mathbf{LT}(f)$ for simplicity.

1.2. Definition Let $\preceq$ be a well-ordering on $\mathcal{B}$.
(i) $\preceq$ is said to be a *monomial ordering* on A if the following conditions are satisfied.

(**MO1**) If $w, u, v, s \in \mathcal{B}$ with $w \prec u$, $\mathbf{LM}(vws) \neq 0$ and $\mathbf{LM}(vus) \neq 0$, then $\mathbf{LM}(vws) \prec \mathbf{LM}(vus)$.

(**MO2**) For $w, u \in \mathcal{B}$, if $u = \mathbf{LM}(vws)$ for some $v, s \in \mathcal{B}$ with $v \neq 1$ or $s \neq 1$, then $w \prec u$. (Hence $1 \prec u$ for all $u \in \mathcal{B}$ with $u \neq 1$.)

(ii) $\preceq$ is said to be a *left monomial ordering* on A if the following conditions are satisfied.

(**LMO1**) If $w, u, v \in \mathcal{B}$ with $w \prec u$, $\mathbf{LM}(vw) \neq 0$ and $\mathbf{LM}(vu) \neq 0$, then $\mathbf{LM}(vw) \prec \mathbf{LM}(vu)$.

(**LMO2**) For $w, u \in \mathcal{B}$, if $u = \mathbf{LM}(vw)$ for some $v \in \mathcal{B}$ with $v \neq 1$, then $w \prec u$. (Hence $1 \prec u$ for all $u \in \mathcal{B}$ with $u \neq 1$.)

(iii) A (left) monomial ordering $\succeq$ on A is called a *graded monomial ordering*, denoted $\succeq_{gr}$, if for $u, w \in \mathcal{B}$ the following holds:

$$u \succ_{gr} w \text{ if and only if } d(u) > d(w) \text{ or } d(u) = d(w) \text{ and } u \succ w.$$

(iv) The triple $(A, \mathcal{B}, \preceq)$ is called an *admissible system* if $\preceq$ is a monomial ordering on A, and is called a *left admissible system* if $\preceq$ is a left monomial ordering on A.

From the definition it is clear that if $(A, \mathcal{B}, \preceq)$ is an admissible system, then it is a left admissible system as well. But if $(A, \mathcal{B}, \preceq)$ is a left admissible system, it seems not necessarily an admissible system since we have the following simple but important fact which is based on the fact that A *may have divisors of zero*. (Nevertheless, we do have many examples of left admissible systems that are admissible systems, e.g., see Example (ii), (iii) below, and later §7.)

1.3.   Lemma Let $(A, \mathcal{B}, \preceq)$ be an admissible system, $f \in A$, where $0 \neq f = \sum_{i=1}^{n} \lambda_i w_i$ such that $\mathbf{LM}(f) = w_1 \succ w_2 \succ \cdots \succ w_n$. For any $u, v \in \mathcal{B}$, if $ufv \neq 0$ then there is some i such that $\mathbf{LM}(ufv) = \mathbf{LM}(uw_iv) \neq 0$ and $\mathbf{LM}(uw_jv) \prec \mathbf{LM}(uw_iv) = \mathbf{LM}(ufv)$ for all $j \neq i$ with $\mathbf{LM}(uw_jv) \neq 0$. (Indeed, all possible nonzero $\mathbf{LM}(uw_jv)$ with $j \neq i$ are among $\mathbf{LM}(uw_{i+\ell}v)$, where $1 \leq \ell \leq n - i$.) Putting $v = 1$ in the above statement we get the similar result for a left admissible system.

$\square$

The monomial w_i appeared in the above lemma will be called the (*left*) *division leading monomial* of f after we distinguish the (left) quasi-zero elements of A in the next section.

Let $I\!N^n$ be the set of all n-tuples $\alpha = (\alpha_1, ..., \alpha_n)$ with $\alpha_i \in I\!N$. Write $|\alpha| = \alpha_1 + \cdots + \alpha_n$. Recall that the *lexicographic ordering* on $I\!N^n$ in natural order, respectively in reverse order, denoted $\geq_{lex}$, is defined as: $\alpha = (\alpha_1, ..., \alpha_n)$, $\beta = (\beta_1, ..., \beta_n) \in I\!N^n$, $\beta >_{lex} \alpha$ if and only if the left-most nonzero entry of $\beta - \alpha = (\beta_1 - \alpha_1, ..., \beta_n - \alpha_n)$ is positive, respectively if and only if the right-most nonzero entry of $\beta - \alpha$ is positive; and the *graded lexicographic ordering* on $I\!N^n$ in either natural order or reverse order, denoted $\geq_{grlex}$, is defined as: $\alpha = (\alpha_1, ..., \alpha_n)$, $\beta = (\beta_1, ..., \beta_n) \in I\!N^n$, $\beta >_{grlex} \alpha$ if and only if

$$|\beta| > |\alpha|, \text{ or } |\beta| = |\alpha| \text{ and } \beta >_{lex} \alpha,$$

where $\geq_{lex}$ is the lexicographic ordering on $I\!N^n$ in either natural order or reverse order. It is well-known (e.g., [BW]) that $\geq_{lex}$ and $\geq_{grlex}$ are well-orderings on $I\!N^n$.

If $A = k[a_1, ..., a_n]$ is a finitely generated k-algebra with n generators $a_1, ..., a_n$, then we call the monomial of the form $a_1^{\alpha_1} \cdots a_n^{\alpha_n}$ a *standard monomial*. Suppose that

$$\mathcal{B} = \left\{ a_1^{\alpha_1} \cdots a_n^{\alpha_n} \mid (\alpha_1, ..., \alpha_n) \in I\!N^n \right\},$$

the set of all standard monomials, forms a k-basis for A. Then it is easy to see that every well-ordering on $I\!N^n$ induces a well-ordering on $\mathcal{B}$, and if $\geq_{lex}$ is the lexicographic ordering on $I\!N^n$ in either natural order or reverse order, then it induces on $\mathcal{B}$ either the monomial ordering $a_1 >_{lex} a_2 >_{lex} \cdots >_{lex} a_n$ or the monomial ordering $a_n >_{lex} a_{n-1} >_{lex} \cdots >_{lex} a_1$. Similarly, the graded lexicographic ordering on $I\!N^n$ induces a graded monomial ordering on $\mathcal{B}$.

It follows from the above remark and CH.I §§3–6 that the following examples are not hard to be verified.

Example (i) Let $A = k[x_1, ..., x_n]$ be the commutative polynomial algebra in n variables and

$$\mathcal{B} = \left\{ x_1^{\alpha_1} \cdots x_n^{\alpha_n} \mid (\alpha_1, ..., \alpha_n) \in I\!N^n \right\}.$$

Then $(A, \mathcal{B}, \geq_{lex})$ and $(A, \mathcal{B}, \geq_{grlex})$ are (left) admissible systems.

(ii) Let $k[x_1, ..., x_n]$ be the commutative polynomial algebra in n variables and

$$A = k[x_1, ..., x_n][x_{n+1}; \sigma_{n+1}, \delta_{n+1}][x_{n+2}; \sigma_{n+2}, \delta_{n+2}] \cdots [x_{n+m}; \sigma_{n+m}, \delta_{n+m}]$$

an iterated skew polynomial k-algebra over $k[x_1, ..., x_n]$, where the σ_{n+j} $(1 \leq j \leq m)$ are algebra endomorphisms and each δ_{n+j} $(1 \leq j \leq m)$ is a σ_{n+j}-derivation, that is, for $f, g \in k[x_1, ..., x_n][x_{n+1}; \sigma_{n+1}, \delta_{n+1}] \cdots [x_{n+j-1}; \sigma_{n+j-1}, \delta_{n+j-1}]$,

$$\delta_{n+j}(fg) = \sigma_{n+j}(f)\delta_{n+j}(g) + \delta_{n+j}(f)g.$$

Then

$$\mathcal{B} = \left\{ x_1^{\alpha_1} \cdots x_n^{\alpha_n} x_{n+1}^{\beta_1} \cdots x_{n+m}^{\beta_m} \;\middle|\; (\alpha_1, ..., \alpha_n, \beta_1, ..., \beta_m) \in \mathbb{N}^{n+m} \right\}$$

forms a k-basis for A. Suppose that for $1 \leq i < j = n + \ell$, $1 \leq \ell \leq m$ we have

$$\sigma_j(x_i) = \lambda_{ij} x_i, \;\; \lambda_{ij} \in k - \{0\},$$
$$\delta_j(x_i) = P_{ij} \in k[x_1, ..., x_n, x_{n+1}, ..., x_{n+h}], \;\; n + h < j = n + \ell.$$

Then $(A, \mathcal{B}, \geq_{lex})$ is a left admissible system, where $\geq_{lex}$ is the lexicographic ordering on $\mathbb{N}^{n+m}$ in the reverse order. If $\deg P_{ij} \leq 2$, then $(A, \mathcal{B}, \geq_{grlex})$ is a left admissible system, where $\geq_{grlex}$ is the graded lexicographic ordering on $\mathbb{N}^{n+m}$ in the reverse order. If $\deg P_{ij} \leq 1$, then $(A, \mathcal{B}, \geq_{grlex})$ is a left admissible system, where $\geq_{grlex}$ is the graded lexicographic ordering on $\mathbb{N}^{n+m}$ in either natural order or reverse order.

Example (ii) above may be applied to CH.I §5 Examples (i), (ii), (iii), (v), (vi), and (vii).

(iii) Let $A = U(\mathfrak{g})$ be the enveloping algebra of an n-dimensional k-Lie algebra $\mathfrak{g} = kx_1 \oplus kx_2 \oplus \cdots \oplus kx_n$ with the bracket multiplication given by

$$(*) \qquad\qquad [x_j, x_i] = \sum_{\ell=1}^{n} \lambda_{ij}^{\ell} x_\ell, \quad j > i.$$

(See CH.I §5.) Then by the PBW theorem,

$$\mathcal{B} = \left\{ x_1^{\alpha_1} \cdots x_n^{\alpha_n} \;\middle|\; (\alpha_1, ..., \alpha_n) \in \mathbb{N}^n \right\}$$

forms a k-basis of A. From the above $(*)$ we may derive that $(A, \mathcal{B}, \geq_{grlex})$ is an admissible system and hence a left admissible system, where $\geq_{grlex}$ is the graded lexicographic ordering on $\mathbb{N}^n$ in either natural order or reverse order.

(iv) Let $A = k\langle X \rangle$ be the free k-algebra on the set $X = \{X_i\}_{i \in \Lambda}$ (see CH.I §1). Then A has the k-basis

$$\mathcal{B} = \left\{ w = X_{i_1} X_{i_2} \cdots X_{i_p} \;\middle|\; X_{i_j} \in X, \; p \geq 1 \right\} \bigcup \{1\}.$$

Recall that the degree of an element $w = X_{i_1} \cdots X_{i_p}$ in $\mathcal{B}$ is defined to be the length p of w, i.e., $d(w) = p$. In particular, $d(1) = 0$. Suppose that Λ is totally ordered by $<$. Recall from [Mor2] that the *graded lexicographic ordering* on $\mathcal{B}$ is defined as: For $u, v \in \mathcal{B}$, $u >_{grlex} v$ if and only if

either $d(u) > d(v)$ or $d(u) = d(v)$ and v is lexicographically less than u,

where v is lexicographically less than u if and only if

either there is $r \in \mathcal{B}$ such that $u = vr$, or there are $X_{j_1}, X_{j_2} \in X$ with $j_1 < j_2$ such that $v = lX_{j_1}r_1, u = lX_{j_2}r_2$ for some $l, r_1, r_2 \in \mathcal{B}$.

Note that $\geq_{grlex}$ induces on X the ordering

$$X_{j_2} >_{grlex} X_{j_1} \text{ whenever } j_2 > j_1 \text{ in } \Lambda.$$

Since $\geq_{grlex}$ is a well-ordering on $\mathcal{B}$, we conclude that $(A, \mathcal{B}, \geq_{grlex})$ is an admissible system and hence a left admissible system.

Observe that all algebras considered above do not have divisors of zero. In §7 we will give examples containing divisors of zero.

Finally, we point out, for a k-algebra A and the fixed k-basis $\mathcal{B}$, that there are many possible monomial orderings on $\mathcal{B}$ and that different orderings lead to different (left) admissible systems.

2. (Left) Quasi-zero Elements and a (Left) Division Algorithm

Let $A = k[a_i]_{i \in \Lambda}$ be a k-algebra with the fixed k-basis

$$\mathcal{B} = \left\{ a_{i_1}^{\alpha_1} \cdots a_{i_n}^{\alpha_n} \mid n \geq 1, \ \alpha_i \in \mathbb{N} \right\}$$

consisting of monomials and 1, and let $\preceq$ be a *well-ordering* on $\mathcal{B}$. In this section, we develop a division algorithm in A, in case $(A, \mathcal{B}, \preceq)$ is a (left) admissible system in the sense of Definition 1.2(iv).

Let $f, g \in A$ be two nonzero elements, where

$$(*) \quad \begin{cases} f = \displaystyle\sum_{i=1}^{m} c_i u_i, \ c_i \in k - \{0\}, \quad u_i \in \mathcal{B}, \ u_1 \succ u_2 \succ \cdots \succ u_m, \\[3mm] g = \displaystyle\sum_{j=1}^{n} \lambda_j w_j, \ \lambda_j \in k - \{0\}, \ w_j \in \mathcal{B}, w_1 \succ w_2 \succ \cdots \succ w_n. \end{cases}$$

If we recall any existed division algorithm from the literature, then roughly speaking, an effective division on f by g should be to "eliminate" the u_i by using certain "monomial multiples" of w_1. Since the k-algebra A *may contain divisors of zero*, in order to have a feasible division algorithm on f by g, it is natural to introduce the following notion and notation.

2.1. Definition Let f be as in $(*)$ above.
(i) If $vfs = 0$ for all $v, s \in \mathcal{B}$ but not both v and s are equal to 1, we say that f is a *quasi-zero element* of A.
If $vf = 0$ for all $v \neq 1$ in $\mathcal{B}$, we say that f is a *left quasi-zero element* of A.
(ii) Put $\mathbf{M}(f) = \{u_i \mid i = 1, ..., m\}$. If f is not a quasi-zero element in the sense of part (i) above, then, with respect to $\preceq$, we set

$$\mathbf{LM}^d(f) = \max \left\{ u_i \in \mathbf{M}(f) \;\middle|\; \begin{array}{l} \text{there are } v, s \in \mathcal{B} \text{ such that } vu_is \neq 0 \\ \text{where either } v \neq 1 \text{ or } s \neq 1 \end{array} \right\}$$

and call $\mathbf{LM}^d(f)$ the *division leading monomial* of f.
If f is not a left quasi-zero element in the sense of part (i) above, then, with respect to $\preceq$, we put

$$\mathbf{LM}^{\ell d}(f) = \max \left\{ u_i \in \mathbf{M}(f) \;\middle|\; \text{there is } v \neq 1 \text{ in } \mathcal{B} \text{ such that } vu_i \neq 0 \right\}$$

and call $\mathbf{LM}^{\ell d}(f)$ the *left division leading monomial* of f.

Let $v, w, s \in \mathcal{B}$ and suppose $vws = \sum_{i=1}^{n} c_i u_i$, where $c_i \in k - \{0\}$, $u_i \in \mathcal{B}$. We may assume that $u_1 \succ u_2 \succ \cdots \succ u_n$. Then $u_1 = \mathbf{LM}(vws)$, and

$$w \prec u_1 \text{ where } \begin{cases} (A, \mathcal{B}, \preceq) \text{ is an admissible system and } v \neq 1 \text{ or } s \neq 1, \\ \text{or} \\ (A, \mathcal{B}, \preceq) \text{ is a left admissible system and } v \neq 1, s = 1. \end{cases}$$

This leads to the divisibility in an admissible system, respectively in a left admissible system.

2.2. Definition Let $(A, \mathcal{B}, \preceq)$ be an admissible system, respectively a left admissible system. For $w, u \in \mathcal{B}$, we say that

u is *divisible* by w if $u = \mathbf{LM}(vws)$ for some $v, s \in \mathcal{B}$,

respectively

u is *divisible by w on left* if $u = \mathbf{LM}(vw)$ for some $v \in \mathcal{B}$.

(Note that the above definition *excludes* the case where vws, respectively vw, is equal to 0.)

Before having a division algorithm in A with respect to a (left) admissible system, several basic facts concerning quasi-zero and non-quasi-zero elements of A are summarized in the following lemma.

2.3. Lemma Let $(A, \mathcal{B}, \preceq)$ be an admissible system (or a left admissible system).
(i) Let $w, u \in \mathcal{B}$. If w is a quasi-zero element, then u is divisible by w if and only if $w = u$.
(ii) If g is a (left) quasi-zero element, $g = \sum_{i=1}^{n} \lambda_i w_i$, then $w_i \neq 1$, $i = 1, ..., n$.
(iii) If $\mathcal{G} = \{g_1, ..., g_s\}$ is any subset of quasi-zero elements of A, then the left and two-sided ideals of A generated by $\mathcal{G}$, respectively, are actually the same. Moreover, the set of all quasi-zero elements of A forms an ideal of A with the zero multiplication on it.
(iv) (a new version of Lemma 1.3) Let f be as in $(*)$ above. If $vfs \neq 0$ for some $v, s \in \mathcal{B}$, where either $v \neq 1$ or $s \neq 1$, then $\mathbf{LM}(vfs) = \mathbf{LM}(v\mathbf{LM}^d(f)s)$. If $vf \neq 0$ for some $v \neq 1$ in $\mathcal{B}$, then $\mathbf{LM}(vf) = \mathbf{LM}(v\mathbf{LM}^{\ell d}(f))$. In the case where $\mathcal{B}$ is closed under multiplication or A is a domain we have $\mathbf{LM}(f) = \mathbf{LM}^d(f) = \mathbf{LM}^{\ell d}(f)$ and hence $\mathbf{LM}(vfs) = \mathbf{LM}(v\mathbf{LM}(f)s)$ and $\mathbf{LM}(vf) = \mathbf{LM}(v\mathbf{LM}(f))$.
□

Given an admissible system $(A, \mathcal{B}, \preceq)$, we now proceed to establish a division algorithm in different cases.

Division on f by g
Input: $f = \sum_{i=1}^{m} c_i u_i$ with $u_1 \succ u_1 \succ \cdots \succ u_m$, and $g = \sum_{j=1}^{n} \lambda_j w_j$ with $w_1 \succ w_2 \succ \cdots \succ w_n$.
Start: If g is a *quasi-zero* element, then clearly none of u_i in $\mathbf{M}(f)$ (Definition 2.1) is divisible by any w_j, $j = 1, ..., n$. In this case we conventionally have
Output 1: write $f = 0 + r$ with $r = f$, and say that *the remainder of f for division by g is r which has the property that none of $u_i \in \mathbf{M}(r) = \mathbf{M}(f)$ is divisible by any $w_j \in \mathbf{M}(g)$.*

If g is *not* a quasi-zero element, we may assume that $\mathbf{LM}^d(g) = w_i$ for some $1 \leq i \leq n$. Note that $\mathbf{LM}(f) = u_1$. If u_1 is divisible by w_i, then

$$u_1 = \mathbf{LM}(vw_is), \quad v, s \in \mathcal{B},$$
(0)   and
$$w_i \prec u_1.$$

Note that since w_i is the division leading monomial of g and $\mathbf{LM}(vw_is) \neq 0$, we have

$$vgs = \lambda_i vw_is + \lambda_{i+1} vw_{i+1}s + \cdots + \lambda_n vw_ns$$

and by Lemma 2.3

$\mathbf{LM}(vw_j s) \prec \mathbf{LM}(vw_i s) = u_1$, whenever $j \neq i$ and $vw_j s \neq 0$,
and
$\dfrac{c_1}{\lambda_i} vgs = c_1 u_1 + g'$ with $g' \in A$, $\mathbf{LM}(g') \prec u_1 = \mathbf{LM}(f)$.

Hence

$$f = \frac{c_1}{\lambda_i} vgs + f_1 \text{ with } f_1 = f - \frac{c_1}{\lambda_i} vgs$$

(1)   and

$$\mathbf{LM}(f_1) \prec u_1 = \mathbf{LM}(f).$$

In the case that u_1 is not divisible by $w_i = \mathbf{LM}^d(g)$, we write

(2) $f = f_1 + r_1$ with $r_1 = c_1 u_1$

and consider the divisibility of $\mathbf{LM}(f_1) = u_2$ by $w_i = \mathbf{LM}^d(g)$.

Since $\preceq$ is a well-ordering on $\mathcal{B}$, after a finite number of repetition of the foregoing process (1)–(2) we arrive at

Output 2: $f = \sum_{\ell=1}^p b_\ell v_\ell g s_\ell + r$, where $b_\ell \in k - \{0\}$, $v_\ell, s_\ell \in \mathcal{B}$, $\mathbf{LM}(v_\ell g s_\ell) \preceq \mathbf{LM}(f)$ whenever $v_\ell g s_\ell \neq 0$, and $r = 0$ or $r = \sum_{k=1}^q c_k t_k$, $c_k \in k - \{0\}$, $t_k \in \mathcal{B}$, with the property that $\mathbf{LM}(r) \preceq \mathbf{LM}(f)$ and none of the t_k is divisible by $w_1, w_2, ..., w_{i-1}, w_i = \mathbf{LM}^d(g)$.

Division on f by $\mathcal{G}$

Input: $f = \sum_{i=1}^m c_i u_i$ with $u_1 \succ u_2 \succ \cdots \succ u_m$, $\mathcal{G} = \{g_1, ..., g_m\} \subset A$, where $0 \neq g_i = \sum_{j=1}^{n_i} \lambda_{ij} w_{ij}$ with $w_{i_1} \succ w_{i_2} \succ \cdots \succ w_{i_{n_i}}$, $i = 1, ..., m$. By the foregoing discussion we may assume that all the g_i are not quasi-zero elements and that $\mathbf{LM}^d(g_i) = w_{ij}$ for some $1 \leq j \leq n_i$, $i = 1, ..., m$.

Start: with $f_0 = f$, $r_0 = 0$.
   If $\mathbf{LM}(f_0)$ is divisible by some $\mathbf{LM}^d(g_i)$, then put

$r_1 = 0,$
$f_1 = f_0 - \dfrac{c_1}{\lambda_i} v_i g_i s_i$, where $\lambda_i \in k - \{0\}$ and $\mathbf{LM}(f_1) \prec \mathbf{LM}(f)$,

and consider the divisibility of $\mathbf{LM}(f_1)$ by $\mathcal{G}$.
   If $\mathbf{LM}(f_0)$ cannot be divided by any $\mathbf{LM}^d(g_i)$, $i = 1, ..., m$, then put

$r_1 = \mathbf{LT}(f_0),$
$f_1 = f_0 - r_1$, where $\mathbf{LM}(f_1) \prec \mathbf{LM}(f_0)$.

Continue the above procedure for f_1 until

Output 3: $f = \sum_{h=1}^d \mu_h v_h g_h s_h + r$, where $\mu_h \in k - \{0\}$, $v_h, s_h \in \mathcal{B}$, $g_h \in \mathcal{G}$, $d \leq m$, $\mathbf{LM}(v_h g_h s_h) \preceq \mathbf{LM}(f)$ whenever $v_h g_h s_h \neq 0$, and $r = 0$ or $r = \sum c_j t_j$, $c_j \in k - \{0\}$, $t_j \in \mathcal{B}$, with the property that $\mathbf{LM}(r) \preceq$

$\mathbf{LM}(f)$ and none of the t_j is divisible by any $w_{i_1}, ..., w_{i_{j-1}}, w_{i_j} = \mathbf{LM}^d(g_i)$, $i = 1, ..., m$.

It is not difficult to see that if $(A, \mathcal{B}, \preceq)$ is a *left* admissible system then a similar division algorithm exists and we have the following

Output 4: Let f and $\mathcal{G}$ be as above. Then $f = \sum_{h=1}^{d} f_h g_h + r$, where $f_h \in A$, $f_h \neq 0$, $g_h \in \mathcal{G}$, $d \leq m$, $\mathbf{LM}(f_h g_h) \preceq \mathbf{LM}(f)$ whenever $f_h g_h \neq 0$, and $r = 0$ or $r = \sum c_j t_j$, $c_j \in k - \{0\}$, $t_j \in \mathcal{B}$, with the property that $\mathbf{LM}(r) \preceq \mathbf{LM}(f)$ and none of the t_j is divisible by any $w_{i_1}, ..., w_{i_{j-1}}, w_{i_j} = \mathbf{LM}^{\ell d}(g_i)$ (Definition 2.2) from left hand side, $i = 1, ..., m$.

2.4. Definition The r appeared in both **Output 3** and **Output 4** above is called a *remainder* of f for division by $\mathcal{G}$, denoted $\overline{f}^{\mathcal{G}}$.

Note that the above $\overline{f}^{\mathcal{G}}$ depends on the order of the g_i in $\mathcal{G}$ and hence it is usually *not unique*. In the next section we will see that if $\mathcal{G}$ is a (left) Gröbner basis in A, then $\overline{f}^{\mathcal{G}}$ is unique and is independent of the order of $g_i's$ in $\mathcal{G}$.

3. (Left) Gröbner Bases in a (Left) Admissible System

Let $(A, \mathcal{B}, \preceq)$ be a (left) admissible system in the sense of §1. Based on the division algorithm developed in §2, in this section we introduce the notion of a (left) Gröbner basis for a (left) ideal in A.

We start with an admissible system $(A, \mathcal{B}, \preceq)$ and a two-sided ideal I of A. For $f \in A$, we may also consider the division on f by I. Maintaining the notation as in §2, after a finite number of division steps on f by the elements in I we arrive at

Output 5: $f = \sum_{i=1}^{d} \lambda_i u_i f_i v_i + r$, where $\lambda_i \in k - \{0\}$, $u_i, v_i \in \mathcal{B}$, $f_i \in I$, $\mathbf{LM}(u_i f_i v_i) \preceq \mathbf{LM}(f)$ whenever $u_i f_i v_i \neq 0$, and $r = 0$ or $r = \sum \mu_j t_j$, $\mu_j \in k - \{0\}$, $t_j \in \mathcal{B}$, with the property that $\mathbf{LM}(r) \preceq \mathbf{LM}(f)$ and none of the t_j is divisible by any $w_1, ..., w_{i-1}, w_i = \mathbf{LM}^d(g)$, where $g = \sum_{i=1}^{n} \lambda_j w_j$ is an arbitrary non-quasi-zero element in I with $\mathbf{LM}^d(g) = w_i$.

Write $\overline{f}^I$ for the remainder r of f for division by I in the above **Output 3**, i.e., $\overline{f}^I = r$. Then it is easy to see that $\overline{f}^I$ is *unique*.

If we start with a left admissible system $(A, \mathcal{B}, \preceq)$, replace I by a left ideal L of A, and use the division on left, then a similar argumentation as given above works for L.

For a subset T in A we write

$$\mathbf{LM}(T) = \left\{ \mathbf{LM}(f) \in \mathcal{B} \mid f \in T \right\}$$
$$\mathbf{O}(T) = \mathcal{B} - \mathbf{LM}(T).$$

3.1. Theorem With notation as above, let T represent an ideal or a left ideal of A. The following holds.

(i) As a k-space, $A = T \oplus \mathrm{Span}_k \mathbf{O}(T)$, where the latter is the k-subspace in A spanned by $\mathbf{O}(T)$.

(ii) There is a k-space isomorphism between A/T and $\mathrm{Span}_k \mathbf{O}(T)$.

(iii) For each $f \in A$, there is a unique $r = \overline{f}^T \in \mathrm{Span}_k \mathbf{O}(T)$ such that $f - r \in T$.

(iv) For $f, g \in A$, $\overline{f}^T = \overline{g}^{'I}$ if and only if $f - g \in T$.

(v) For $f \in A$, $f \in T$ if and only if $\overline{f}^T = 0$.

Proof We prove the theorem for the two-sided ideal T because a similar argumentation works for left ideals.

(i) Let f be an element of A. By the division algorithm we have $f = \sum_{i=1}^{d} \lambda_i v_i f_i s_i + r$, where $\lambda_i \in k - \{0\}$, $v_i, s_i \in \mathcal{B}$, $f_i \in T$, $\mathbf{LM}(v_i f_i s_i) \preceq \mathbf{LM}(f)$ whenever $v_i f_i s_i \neq 0$, and $r = \overline{f}^T$. If $r \neq 0$, we may assume that $r = \sum_\ell b_\ell t_\ell$, where $b_\ell \in k - \{0\}$ and $t_\ell \in \mathcal{B}$. Since r has the property that none of the t_ℓ is divisible by any $w_1, ..., w_{i-1}, w_i = \mathbf{LM}^d(g)$, where $g = \sum_{h=1}^{m} \lambda_h w_h$ is an arbitrary non-quasi-zero element in T, it follows that none of the t_ℓ belongs to $\mathbf{LM}(T)$. This proves that $r \in \mathrm{Span}_k \mathbf{O}(T)$ and hence $f \in T + \mathrm{Span}_k \mathbf{O}(T)$. That $T \cap \mathrm{Span}_k \mathbf{O}(T) = \{0\}$ is clear by the definition of $\mathbf{O}(T)$. Therefore, $A = T \oplus \mathrm{Span}_k \mathbf{O}(T)$.

(ii) By (i) above, this is clear.

(iii) Take $f \in A$ and consider the division of f by T. If $f = \sum \lambda_i v_i f_i s_i + r = \sum \lambda_j w_j f_j u_j + r'$, namely there are two remainders r and r' for the division of f by T, then it follows from the proof of (i) that $r - r' \in T \cap \mathrm{Span}_k \mathbf{O}(T) = \{0\}$. Hence $r = r'$, as desired.

(iv) and (v) follow from the foregoing (i)–(iii).   □

Considering a generating set for an ideal I, respectively for a left ideal L, the above discussion leads the definition of a Gröbner basis for I, respectively a left Gröbner basis for L.

3.2. Definition Let $(A, \mathcal{B}, \preceq)$ be a (left) admissible system, and $\mathcal{G} = \{g_i\}_{i \in J} \subset A$.

(i) For the two-sided ideal of A generated by $\mathcal{G}$, denoted $I = \langle \mathcal{G} \rangle$, if every $f \in I$ has a presentation

$$f = \sum_i \lambda_i u_i g_i v_i, \quad \lambda_i \in k - \{0\}, \ u_i, v_i \in \mathcal{B}, \ g_i \in \mathcal{G}$$

such that $\mathbf{LM}(u_i g_i v_i) \preceq \mathbf{LM}(f)$ whenever $u_i g_i v_i \neq 0$, then $\mathcal{G}$ is called a *Gröbner basis* of I.

(ii) For the left ideal of A generated by $\mathcal{G}$, denoted $L = \langle \mathcal{G}]$, if every $f \in L$ has a presentation

$$f = \sum_j h_j g_j, \quad \lambda_j \in k - \{0\}, \ h_j \in A, \ g_j \in \mathcal{G}$$

such that $\mathbf{LM}(h_j g_j) \preceq \mathbf{LM}(f)$ whenever $h_j g_j \neq 0$, then $\mathcal{G}$ is called a *left Gröbner basis* of L.

The presentation for f in (i), respectively in (ii) above, is called a *Gröbner representation* of f, respectively a *left Gröbner representatin* of f by $\mathcal{G}$.

(iii) A (left) Gröbner basis $\mathcal{G}$ is called *reduced* if, for all j, $\mathbf{LC}(g_j) = 1$ and no nonzero term in g_j is divisible by any $\mathbf{LM}(g_i)$ for any $j \neq i$.

Remark (i) In the above definition the reason that we do not reqiure $\mathcal{G}$ to be finite is obvious, namely, *not every A is (left) Noetherian*, and consequently, not every (left) ideal in A is finitely generated. Moreover, from [Mor2] or [Ufn] (or the example given in the next section) one may see that, even if we start with a single element, a noncommutative version of Buchberger's algorithm may produce an infinite Gröbner basis.

(ii) At this stage, one might be wondering the case where $\mathcal{G} = \{g_i\}_{i \in J}$ consists of quasi-zero elements. Indeed, it easily follows from Lemma 2.3 that $\mathcal{G}$ is a (left) Gröbner basis for the (left) ideal of A generated by $\mathcal{G}$. Moreover, let I, respectively L, be the ideal, respectively the left ideal of A, generated by the subset $\mathcal{G} = \{g_i\}_{i \in J} \cup \{g_k\}_{k \in T}$ where $\{g_i\}_{i \in J}$ consists of quasi-zero elements and $\{g_k\}_{k \in T}$ does not contain quasi-zero element. Suppose that $\mathcal{G}$ is a Gröbner basis of I, respectively a left Gröbner basis of L. Then for any $f \in I$, respectively $f \in L$ but $f \notin \{g_i\}_{i \in J}$, a Gröbner representation of f is completely determined by $\{g_k\}_{k \in T}$.

Theorem 3.1 now makes the algorithmic feature of a (left) Gröbner basis more clearer.

3.3. Proposition Let $\mathcal{G} = \{g_i\}_{i \in J}$ be a Gröbner basis for an ideal I, or a left Gröbner basis for a left ideal L. Let $f \in A$ and $\overline{f}^{\mathcal{G}}$ the remainder of f for division by $\mathcal{G}$.

(i) $\overline{f}^{\mathcal{G}}$ is independent of the order of the $g_i \in \mathcal{G}$, and $\overline{f}^{\mathcal{G}}$ is unique in A.

(ii) $f \in I$ or $f \in L$ if and only if $\overline{f}^{\mathcal{G}} = 0$.

□

3.4. Proposition Let $\mathcal{G} = \{g_i\}_{i \in J}$ be a subset of A and I, respectively L, the two-sided ideal of A, respectively the left ideal of A, generated by $\mathcal{G}$. Then $\mathcal{G}$ is a Gröbner basis of I, respectively a left Gröbner basis of L if and only if $\overline{f}^{\mathcal{G}} = 0$ for every $f \in I$, respectively for every $f \in L$.

□

It is easy to see that Definition 3.2 includes all well-known definitions of (left) Gröbner bases in the literature.

Historically, the successful development of commutative Gröbner basis theory was motivated by a study of the problem in determining the k-basis of the vector space $k[x_1, ..., x_n]/I$ by using a (finite) generating set of I, where I is an ideal of the commutative polynomial algebra $k[x_1, ..., x_n]$ over a field k (see [Eis] 15.6, History of Gröbner Bases). Theorem 3.1 shows, not exceptionally, that various noncommutative Gröbner basis theories fully share this feature as a starting point. The algorithmic and structural aspects of Theorem 3.1 will become more apparent in CH.III when we apply the very noncommutative Gröbner bases of ([Berg], [Mor1–2]) to algebras defined by relations.

4. (Left) Gröbner Bases in Various Contexts

In the last section, (left) Gröbner bases are defined in a quite general extent. Since various noncommutativities lead to various noncommutative algebraic structures, from a practical point of view, it is then necessary to indicate the main differences of (left) Gröbner bases in various contexts.

Let I be a two-sided ideal in one of the following algebras:

$A = k[x_1, ..., x_n]$, the commutative polynomial algebra in $x_1, ..., x_n$,
$A = k\langle X_1, ..., X_n \rangle$, the noncommutative free algebra in $X_1, ..., X_n$.

Suppose that $(A, \mathcal{B}, \preceq)$ is an admissible system with $\mathcal{B}$ the standard k-basis. Then the following results are well known.

(i) A subset $\mathcal{G} = \{g_i\}_{i \in J}$ is a Gröbner basis of I if and only if $\langle \mathbf{LM}(f) \mid f \in I \rangle = \langle \mathbf{LM}(g_i) \mid g_i \in \mathcal{G} \rangle$. (Indeed the latter property has been used as the definition of a Gröbner basis for I, e.g., [BW], [CLO'], [Mor2]. In [Gr] this is also used as the definition of a Gröbner basis for a two-sided ideal in an algebra, possibly with no 1 but possibly with divisors of zero.)

(ii) If $\mathcal{G} = \{w_i\}_{i \in J} \subset \mathcal{B}$ and I is the ideal generated by $\mathcal{G}$, then $\mathcal{G}$ is a Gröbner basis of I.

Note that the above (i)–(ii) are based on the fact that

- the k-basis $\mathcal{B}$ used in those cases is closed under multiplication.

For an arbitrary associative algebra A, however, the above property ($\bullet$) no longer necessarily exists. The following 4.1–4.2 brings us both good and bad news on the impact of this fact more precisely in a general setting.

4.1. Proposition Let $(A, \mathcal{B}, \preceq)$ be an admissible system, and let I, respectively L, be an ideal, respectively a left ideal, generated by $\mathcal{G} = \{w_i\}_{i \in J} \subset \mathcal{B}$. Suppose $\mathcal{B}$ satisfies:

$$(*) \qquad\qquad w, v \in \mathcal{B} \text{ implies } \lambda w v \in \mathcal{B} \text{ for some } \lambda \in k.$$

Then $\mathcal{G}$ is a Gröbner basis for I, respectively a left Gröbner basis for L. A similar statement holds for a left admissible system and any left ideal L.

If $\mathcal{B}$ does not satisfy the condition $(*)$ above, then $\mathcal{G}$ is not necessarily a Gröbner basis for I, respectively not necessarily a left Gröbner basis for L.

Proof Let $I = \langle \mathcal{G} \rangle$ with $\mathcal{G} = \{w_i\}_{i \in J}$. If $f \in I$ we may assume that $f = \sum_{i=1}^{s} h_i w_i g_i$ for some $h_i, g_i \in A$ and that

$$w_1 \succ w_2 \succ \cdots \succ w_s,$$
$$h_i = \sum_{j}^{n_i} \lambda_{i_j} t_{i_j}, \ t_{i_1} \succ t_{i_2} \succ \cdots \succ t_{i_{n_i}},$$
$$g_i = \sum_{d}^{m_i} \lambda_{i_d} v_{i_d}, \ v_{i_1} \succ v_{i_2} \succ \cdots \succ v_{i_{m_i}}.$$

Thus, $\mathbf{LM}(h_i) = t_{i_1}$ and $\mathbf{LM}(g_i) = v_{i_1}$. By the assumption $(*)$ on $\mathcal{B}$, there are $\lambda_i \in k$, $i = 1, ..., s$, such that

$$\mathbf{LM}(h_i w_i g_i) = \lambda_i t_{i_1} w_i v_{i_1}, \quad i = 1, ..., s.$$

Since for $i = 1, ..., s$ we have

$$t_{i_1} w_i v_{i_1} \succ t_{i_1} w_{i+\ell} v_{i_1} \succ t_{(i+\ell)_1} w_{i+\ell} v_{i_1} \succ t_{(i+\ell)_1} w_{i+\ell} v_{(i+\ell)_1},$$

it follows that $t_{i_1} w_i v_{i_1} \neq t_{(i+\ell)_1} w_{i+\ell} v_{(i+\ell)_1}$, $i = 1, ..., s-1$, $\ell = 1, ..., s-i$. Thus, $\mathbf{LM}(h_i w_i g_i) \preceq \mathbf{LM}(f)$ whenever $h_i w_i g_i \neq 0$. This shows that f has a Gröbner representation. Hence, $\{w_i\}_{i \in J}$ is a Gröbner basis for I. A similar argumentation works for L.

Now let us consider the first Weyl algebra $A_1(k) = k[x, \partial]$ (CH.I §5) for finding a counterexample in the case where $\mathcal{B}$ does not have the property $(*)$. First note that A is a solvable polynomial algebra with respect to $x >_{grlex} \partial$ (see later §7). Let $L = \langle g_1, g_2 \rangle$ be the left ideal generated by the standard monomials $g_1 = x$, $g_2 = \partial^2$ in A. An easy calculation shows that the S-element of g_1 and g_2

in A, denoted $\mathrm{S}(g_1,\ g_2)$ (see §7 for the definition), equals 2∂ which can neither be divided by $\mathbf{LM}(g_1) = x$ nor be divided by $\mathbf{LM}(g_2) = \partial^2$. Hence $\{g_1,\ g_2\}$ is not a left Gröbner basis for L (see §5 Theorem 5.1 below).

We leave the counterexample for a two-sided ideal in A to the reader. (Hint: Consider $I = \langle x, \partial \rangle$.) □

In the above proposition, if $(A, \mathcal{B}, \preceq)$ is *only* a left admissible system and $\mathcal{B}$ does not have the property $(*)$, then we are sure that $\mathcal{G}$ is not necessarily a left Gröbner basis for $L = \langle \mathcal{G} \rangle$. But the author failed to have a counterexample.

4.2. Proposition (i) Let $(A, \mathcal{B}, \preceq)$ be an admissible system, and let I be an ideal of A. If a subset $\mathcal{G} = \{g_i\}_{i \in J} \subset I$ is such that $\langle \mathbf{LM}(f) \mid f \in I \rangle = \langle \mathbf{LM}(g_i) \mid g_i \in \mathcal{G} \rangle$, then $\mathcal{G}$ is a Gröbner basis for I.

(ii) Let $(A, \mathcal{B}, \preceq)$ be a left admissible system, and let L be a left ideal of A. If a subset $\mathcal{G} = \{g_i\}_{i \in J} \subset L$ is such that $\langle \mathbf{LM}(f) \mid f \in L] = \langle \mathbf{LM}(g_i) \mid g_i \in \mathcal{G}]$, then $\mathcal{G}$ is a left Gröbner basis for L.

The converse of the above (i) and (ii) is not necessarily true.

Proof We only give the proof for (i) and a similar argumentation will do for (ii). We prove that under the assumption of (i) every element $f \in I$ has a Gröbner representation by $\mathcal{G}$. To see this, consider the division of f by $\mathcal{G}$. Then we have $f = \sum v_i g_i s_i + r$, where $v_i, s_i \in \mathcal{B}$, $\mathbf{LM}(v_i g_i s_i) \preceq \mathbf{LM}(f)i$ whenever $v_i g_i s_i \neq 0$, and $r = \overline{f}^{\mathcal{G}}$. If $r \neq 0$, then since $r \in I$ and $0 \neq \mathbf{LM}(r) \in \langle \mathbf{LM}(f) \mid f \subset I \rangle = \langle \mathbf{LM}(g_i) \mid g_i \in \mathcal{G} \rangle$. Thus, $\mathbf{LM}(r) = \sum_{i,j} w_{i_j} \mathbf{LM}^d(g_i) u_{i_j}$ with $w_{i_j}, u_{i_j} \in \mathcal{B}$ and hence $\mathbf{LM}(r) = \mathbf{LM}(w_{i_j} \mathbf{LM}^d(g_i) u_{i_j})$ for some g_i. But this contradicts that $r = \overline{f}^{\mathcal{G}}$. Therefore $r = 0$, i.e., f has a Gröbner representation by $\mathcal{G}$, as desired.

To see why the converse of (i) and (ii) are not necessarily true, again, let us consider the first Weyl algebra $A_1(k) = k[x, \partial]$ for finding a counterexample for (ii) and leave the counterexample for (i) to the reader. Let $L = \langle g \rangle$ be the left ideal generated by $g = x + 1$ in A. Using the monomial ordering $x >_{grlex} \partial$, it follows from later §5 Theorem 5.1 and §7 Proposition 7.4 that $\mathcal{G} = \{g = x + 1\}$ is a left Gröbner basis for L and $\mathbf{LM}(g) = x$. Now $\partial g = x\partial + \partial + 1 = h \in L$ implies that $\mathbf{LM}(h) = x\partial \in \mathbf{LM}(L)$. Suppose $x\partial = fx$ for some $f \in A$. Then since it is easy to see that $f \circ x = 0$, i.e., the action of the operator f on x is zero, we have $f = \sum_{\beta_i \geq 2} \lambda_i x^{\alpha_i} \partial^{\beta_i}$. Thus

$$
\begin{aligned}
x\partial = fx &= \left(\sum_{\beta_i \geq 2} \lambda_i x^{\alpha_i} \partial^{\beta_i} \right) x \\
&= \sum_{\beta_i \geq 2} \lambda_i x^{\alpha_i} (x\partial^{\beta_i} + \beta_i \partial^{\beta_i - 1}) \\
&= \sum_{\beta_i \geq 2} \lambda_i x^{\alpha_i + 1} \partial^{\beta_i} + \sum_{\beta_i \geq 2} \lambda_i \beta_i x^{\alpha_i} \partial^{\beta_i - 1}.
\end{aligned}
$$

Note that $\{x^\alpha \partial^\beta \mid (\alpha, \beta) \in I\!N^2\}$ is a k-basis for A. It follows that $f = 0$. This proves that $\mathbf{LM}(h) = x\partial \notin \langle \mathbf{LM}(g)] = \langle x]$.   □

Another typical feature of Gröbner bases in a commutative polynomial algebra is that any *single* polynomial forms a Gröbner basis for the principal ideal generated by that polynomial; and this is even true for (noncommutative) solvable polynomial algebras, as one may easily verify by definitions (see §7). The example given below shows that this nice property does not exist in general (also see [Ufn]).

Example (i) Let $A = k\langle X, Y \rangle$ be the free algebra in two variables X and Y, and let $(A, \mathcal{B}, \geq_{grlex})$ be the admissible system as given in §1, where $X >_{grlex} Y$. Consider the ideal $I = \langle g = X^2 - XY \rangle$ generated by the single element $g = X^2 - XY$. Then we claim that
(a) $\langle X^2 - XY \rangle = I = \langle XY^n X - XY^{n+1} \rangle_{n \geq 0}$.
(b) $\{g = X^2 - XY\}$ is not a Gröbner basis for I.
(c) $\mathcal{G} = \{g_n = XY^n X - XY^{n+1}\}_{n \geq 0}$ is an infinite Gröbner basis for I.

Proof (a) We do induction on n. If $n = 0$, then $g_0 = XY^0 X - XY^{0+1} = X^2 - XY = g \in I$. Hence $I \subset \langle XY^n X - XY^{n+1} \rangle$. Suppose $XY^{n-1}X - XY^n = g_{n-1} \in I$ for $n \geq 1$. We prove that $XY^n X - XY^{n+1} = g_n \in I$. Note that $X^2 - XY = g \in I$ implies both $X^2 Y^n - XY^{n+1}$ and $-X^2 Y^{n-1}X + XY^n X$ are in I. It follows that

$$-X(XY^{n-1}X - XY^n) + (XY^n X - XY^{n+1}) \in I.$$

By the induction hypothesis, $XY^n X - XY^{n+1} = g_n \in I$, as desired.
(b) From part (a) we know that $XYX - XY^2 = g_1 \in I$ with $\mathbf{LM}(g_1) = XYX$. Note that $\mathbf{LM}(g) = X^2$. It is clear that g_1 cannot have a Gröbner representation by g.
(c) This can be verified either by a programme of [FG], or by an algorithm of [Mor2], or by directly checking that every S-element $S(g_i, g_j)$ has a Gröbner representation by $\mathcal{G}$ (see §5 blow).   □

Remark It is an interesting exercise to check what the result will be in the above example if the ordering $Y >_{grlex} X$ is used. (Also see CH.III Corollary 1.3.)

5. (Left) S-elements and Buchberger Theorem

After giving the definition and deriving some basic properties for (left) Gröbner bases in (left) admissible   systems,   one may also expect to see the possibility of having a version of Buchberger's algorithm in order to produce a (left) Gröbner basis in a given (left) admissible system. Since our definition of (left) Gröbner basis is compatible with all existed definitions, knowing from the literature, this is only a matter of learning the mathematics principle behind Buchberger's algorithm in the commutative case, carefully.

Let $A = k[x_1, ..., x_n]$ be the commutative polynomial k-algebra in n variables and $(A, \mathcal{B} \preceq)$ an admissible system as given in §1 Example (i). Let $\alpha = (\alpha_1, ..., \alpha_n), \beta = (\beta_1, ..., \beta_n) \in I\!N^n$ and $f, g \in A$ with $\mathbf{LM}(f) = x^\alpha = x_1^{\alpha_1} \cdots x_n^{\alpha_n}$, $\mathbf{LM}(g) = x^\beta = x_1^{\beta_1} \cdots x_n^{\beta_n}$. Put

$$\gamma = (\gamma_1, ..., \gamma_n) \text{ with } \gamma_i = \max\{\alpha_i, \beta_i\}.$$

Then

$(*)$ $$x^{\gamma - \alpha} x^\alpha = x^{\gamma - \beta} x^\beta$$

and the S-*polynomial* of f and g is defined as:

$$S(f, g) = \frac{x^{\gamma - \alpha}}{\mathbf{LC}(f)} f - \frac{x^{\gamma - \beta}}{\mathbf{LC}(g)} g.$$

Theorem (Buchberger) Let $\mathcal{F} = \{f_1, ..., f_s\}$ be a generating set of the ideal $I \subset A$. Then $\mathcal{F}$ is a Gröbner basis for I if and only if $\overline{S(f_i, f_j)}^{\mathcal{F}} = 0$ for all $f_i \neq f_j \in \mathcal{F}$.

Proof We adopt the proof of this theorem given in [AdL], and first prove a preliminary result.

• Let $f_1, ..., f_s \in k[x_1, ..., x_n]$ be such that $\mathbf{LM}(f_i) = x^\alpha = x_1^{\alpha_1} \cdots x_n^{\alpha_n} \neq 0$ for all $i = 1, ..., s$. Let $f = \sum_{i=1}^s c_i f_i$ with $c_i \in k$. If $\mathbf{LM}(f) \prec x^\alpha$, then f is a k-linear combination of $S(f_i, f_j)$, $1 \leq i < j \leq s$.

Write $f_i = a_i x^\alpha +$ lower terms, $a_i \in k$. Then the assumption yields that $\sum_{i=1}^s c_i a_i = 0$. Since $\mathbf{LM}(f_i) = \mathbf{LM}(f_j)$, it follows that $S(f_i, f_j) = \frac{1}{a_i} f_i - \frac{1}{a_j} f_j$. Thus

$$
\begin{aligned}
f &= c_1 f_1 + \cdots + c_s f_s \\
&= c_1 a_1 \left(\frac{1}{a_1} f_1 \right) + \cdots + c_s a_s \left(\frac{1}{a_s} f_s \right) \\
&= c_1 a_1 \left(\frac{1}{a_1} f_1 - \frac{1}{a_2} f_2 \right) + (c_1 a_1 + c_2 a_2) \left(\frac{1}{a_2} f_2 - \frac{1}{a_3} f_3 \right) + \cdots \\
&\quad + (c_1 a_1 + \cdots + c_{s-1} a_{s-1}) \left(\frac{1}{a_{s-1}} f_{s-1} - \frac{1}{a_s} f_s \right) + (c_1 a_1 + \cdots + c_s a_s) \frac{1}{a_s} f_s \\
&= c_1 a_1 S(f_1, f_2) + (c_1 a_1 + c_2 a_2) S(f_2, f_3) + \cdots \\
&\quad + (c_1 a_1 + \cdots + c_{s-1} a_{s-1}) S(f_{s-1}, f_s),
\end{aligned}
$$

because $c_1 a_1 + \cdots + c_s a_s = 0$. This proves the conclusion ($\bullet$).

We are now ready to prove Buchberger theorem.

If $\mathcal{F}$ is a Gröbner basis for I, then $\overline{S(f_i, f_j)}^{\mathcal{F}} = 0$ for all $i \neq j$ by Proposition 3.4 since $S(f_i, f_j) \in I$.

Conversely, let us assume that $\overline{S(f_i, f_j)}^{\mathcal{F}} = 0$ for all $i \neq j$. We will prove that every $f \in I$ has a Gröbner representation with respect to $\mathcal{F}$. Write $f = \sum_{i=1}^{s} h_i f_i$ and put

$$x^\alpha = x_1^{\alpha_1} \cdots x_n^{\alpha_n} = \max \left\{ \mathbf{LM}(h_i f_i) \,\Big|\, 1 \le i \le s \right\}.$$

If $x^\alpha \preceq \mathbf{LM}(f)$, we are done. Otherwise, suppose $\mathbf{LM}(f) \prec x^\alpha$ and let

$$T = \left\{ i \,\Big|\, \mathbf{LM}(h_i f_i) = x^\alpha \right\}.$$

For $i \in T$, write $h_i = c_i \mathbf{LM}(h_i) +$ lower terms. Set $g = \sum_{i \in T} c_i \mathbf{LM}(h_i) f_i$. Then $\mathbf{LM}(\mathbf{LM}(h_i) f_i) = x^\alpha$ for all $i \in T$, but $\mathbf{LM}(g) \prec x^\alpha$. By the preliminary result obtained above, there exists $d_{ij} \in k$ such that

$$g = \sum_{i,j \in T, i \neq j} d_{ij} S(\mathbf{LM}(h_i) f_i, \mathbf{LM}(h_j) f_j).$$

Now, $x^\alpha = \mathrm{lcm}\left(\mathbf{LM}(\mathbf{LM}(h_i) f_i), \mathbf{LM}(\mathbf{LM}(h_j) f_j) \right)$, so

$$
\begin{aligned}
S\left(\mathbf{LM}(h_i) f_i, \mathbf{LM}(h_j) f_j \right) &= \frac{x^\alpha}{\mathbf{LT}(\mathbf{LM}(h_i) f_i)} \mathbf{LM}(h_i) f_i - \\
&\qquad - \frac{x^\alpha}{\mathbf{LT}(\mathbf{LM}(h_j) f_j)} \mathbf{LM}(h_j) f_j \\
&= \frac{x^\alpha}{\mathbf{LT}(f_i)} f_i - \frac{x^\alpha}{\mathbf{LT}(f_j)} f_j \\
&= \frac{x^\alpha}{x^\beta} S(f_i, f_j),
\end{aligned}
$$

where $x^\beta = \mathrm{lcm}(\mathbf{LM}(f_i), \mathbf{LM}(f_j))$. By the assumption, $\overline{S(f_i, f_j)}^{\mathcal{F}} = 0$, and hence it is not hard to see from this last equation that

$$\overline{S(\mathbf{LM}(h_i) f_i, \mathbf{LM}(h_j) f_j)}^{\mathcal{F}} = 0.$$

This yields a representation

$$S(\mathbf{LM}(h_i) f_i, \mathbf{LM}(h_j) f_j) = \sum_{\ell=1}^{s} h_{ij\ell} f_\ell,$$

where by the division algorithm

$$
\begin{aligned}
\max \left\{ \mathbf{LM}(h_{ij\ell} f_\ell) \,\Big|\, 1 \le \ell \le s \right\} &= \mathbf{LM}(S(\mathbf{LM}(h_i) f_i, \mathbf{LM}(h_j) f_j)) \\
&\prec \max \{ \mathbf{LM}(\mathbf{LM}(h_i f_i), \mathbf{LM}(\mathbf{LM}(h_j f_j) \} \\
&= x^\alpha.
\end{aligned}
$$

Substituting these expressions into g above, and g into f, we get $f = \sum_{i=1}^{s} h_i' f_i$, with $\max\{\mathbf{LM}(h_i' f_i) \mid 1 \le i \le s\} < x^{\alpha}$. This is a contradiction. $\square$

Remark From a structural point of view, the above proof indeed tells us that

- Buchberger theorem is equivalent to: $\mathcal{F}$ is a Gröbner basis if and only if every $\mathrm{S}(f_i, f_j)$ has a Gröbner representation by $\mathcal{F}$.

The following algorithm tests whether $\mathcal{F}$ is a Gröbner basis or not; if not, it produces a finite Gröbner basis containing $\mathcal{F}$.

Buchberger's algorithm

> **Input:** $\mathcal{F} = \{f_1, ..., f_s\}$
> **Output:** $\mathcal{G} = \{g_1, ..., g_m\}$, a Gröbner basis for I
> **Initialization:** $\mathcal{G} := \mathcal{F}$, $\mathcal{S} := \{\{f_i, f_j\} \mid f_i \ne f_j \in \mathcal{F}\}$
> **While** $\mathcal{S} \ne \emptyset$ **Do**
> $\quad$ Choose any $\{f_i, f_j\} \in \mathcal{S}$
> $\quad$ $\mathcal{S} := \mathcal{S} - \{\{f_i, f_j\}\}$
> $\quad$ $\overline{\mathrm{S}(f_i, f_j)}^{\mathcal{G}} = r$
> $\quad$ **If** $r \ne 0$ **Then**
> $\quad\quad$ $\mathcal{S} := \mathcal{S} \cup \{\{u, r\} \mid \text{for all } u \in \mathcal{G}\}$
> $\quad\quad$ $\mathcal{G} := \mathcal{G} \cup \{r\}$

Although it is a well-know fact in the commutative case, it is necessary to see why this algorithm terminates in a finite steps and then the output result gives us a finite Gröbner basis for I.

First suppose to the contrary that the algorithm does not terminate. Then, as the algorithm progresses, we construct a set $\mathcal{G}_i$ strictly larger than $\mathcal{G}_{i-1}$ and obtain a strictly increasing infinite sequence

$$\mathcal{G}_1 \subset \mathcal{G}_2 \subset \mathcal{G}_3 \subset \cdots.$$

Each $\mathcal{G}_i$ is obtained from $\mathcal{G}_{i-1}$ by adding some $r \in I$ to $\mathcal{G}_{i-1}$, where r is the nonzero remainder, with respect to $\mathcal{G}_{i-1}$, of an S-polynomial of two elements of $\mathcal{G}_{i-1}$. By the property of the remainder we know that $\mathbf{LM}(r)$ cannot be divided by any of the $\mathbf{LM}(f)$ with $f \in \mathcal{G}_{i-1}$. Writing $\langle \mathbf{LM}(\mathcal{G}_i) \rangle$ for the ideal generated by the $\mathbf{LM}(f)$ with $f \in \mathcal{G}_i$, we get

$$\langle \mathbf{LM}(\mathcal{G}_1) \rangle \subset \langle \mathbf{LM}(\mathcal{G}_2) \rangle \subset \langle \mathbf{LM}(\mathcal{G}_3) \rangle \subset \cdots.$$

This is a strictly asccending chain of ideals in A which contradicts the Noetherian property of A.

Now we have $\mathcal{F} \subseteq \mathcal{G} \subseteq I$, and hence $I = \langle f_1, ..., f_s \rangle \subseteq \langle g_1, ..., g_t \rangle \subseteq I$. Thus $\mathcal{G}$ is a generating set for the ideal I. Moreover, if g_i, g_j are polynomials in $\mathcal{G}$, then $\overline{\mathrm{S}(g_i, g_j)}^{\mathcal{G}} = 0$ by construction. Therefore $\mathcal{G}$ is a Gröbner basis for I by Buchberger theorem. $\square$

What did we learn so far?

(i) To have a noncommutative version of Buchberger's algorithm for a given (left) admissible system, it is then clear that we have to first look for an appropriate analogue of the S-polynomial. For a given admissible system $(A, \mathcal{B}, \preceq)$, where $A = [a_i]_{i \in \Lambda}$, and for $f, g \in A$ with

$$\mathbf{LM}^d(f) = a_{i_1}^{\alpha_1} \cdots a_{i_n}^{\alpha_n} = w, \quad \mathbf{LM}^g(g) = a_{j_1}^{\beta_1} \cdots a_{j_m}^{\beta_m} = u \in \mathcal{B},$$

we say that *an S-element* about f and g, denoted $S(f, g)$, is *constructable* provided f and g are *not quasi-zero* elements (see §3 Remark(ii) for the reason) and there are $s, r, v, l \in \mathcal{B}$ such that

$$\mathbf{LM}(s\mathbf{LM}^d(f)r) = \mathbf{LM}(v\mathbf{LM}^d(g)l) = 0,$$

or

$$\mathbf{LM}(s\mathbf{LM}^d(f)r) = \mathbf{LM}(v\mathbf{LM}^d(g)l) \neq 0.$$

If it is the case, then we write

$$S(f, g) = sfr - vgl \text{ or } S(f, g) = \frac{\mu}{\lambda} sfr - vgl$$

where $\lambda = \mathbf{LC}(s(f)r)$ and $\mu = \mathbf{LC}(v(g)l)$ in the "or" case.

Note that if $S(f, g)$ is constructable, then the quaternion (s, r, v, l) *may not be unique*, or in other words, taking a unique "l.c.m" (as in the commutative case) is not always possible, that is, all "overlaps" of w and u must be considered in order to construct *all* S-elements about f, g (e.g., see [Gr], [Mor2]). This may be illustrated by looking at $A = k\langle X \rangle$, the free algebra on $X = \{X_i\}_{i \in \Lambda}$ with $\mathcal{B}$ its standard k-basis. For $u, w \in \mathcal{B}$, there are two possible relations that should be considered:

(a) If u and w have proper overlaps, then $sur = vwl$ for some $s, r, v, l \in \mathcal{B}$;
(b) If (a) is excluded, then trivially $1uw = uw1$, $wu1 = 1wu$.

For instance, if $w = X_1 X_2 X_3 X_2 X_3$ and $u = X_3 X_2 X_3 X_4$, we must consider both $X_1 X_2 (X_3 X_2 X_3)$ with $(X_3 X_2 X_3) X_4$ as one overlap and $X_1 X_2 X_3 X_2 (X_3)$ with $(X_3) X_2 X_3 X_4$ as another overlap. Now use $\geq_{gr}$ on $\mathcal{B}$. Then, for $f = w + X_1^2 + 1$ and $g = u + 5 X_2 X_3 X_1 + 2 X_3 X_2$, we have $\mathbf{LM}^d(f) = \mathbf{LM}(f) = w$, $\mathbf{LM}^d(g) = \mathbf{LM}(g) = u$, and

$$S_1(f, g) = fX_4 - X_1 X_2 g,$$
$$S_2(f, g) = fX_2 X_3 X_4 - X_1 X_2 X_3 X_2 g.$$

A similar argumentation works for constructing *left* S-elements in a *left* admissible system.

(ii) We also note that the idea of using S-elements in Buchberger theorem is indeed not restricted to finite set of generators. More precisely, a version of Buchberger theorem may always be mentioned as follows.

5.1. Theorem (noncommutative version of Buchberger theorem) Let $(A, \mathcal{B}, \preceq)$ be a (left) admissible system and $\mathcal{G} = \{g_i\}_{i \in J}$ a subset of $\quad$ *non-quasi-zero* elements in A. Suppose that, for each pair (g_i, g_j), a (left) S-element $S(g_i, g_j)$ is constructable, $i, j \in J$. Then $\mathcal{G}$ is a (left) Gröbner basis in A if and only if *every* (left) S-element $S(g_i, g_j)$ has a (left) Gröbner representation by $\mathcal{G}$, or in other words, if and only if $\overline{S(g_i, g_j)}^{\mathcal{G}} = 0$.

$\square$

(iii) Finally, for a given finite subset $\mathcal{F} = \{f_1, ..., f_s\}$ in a (left) admissible system $(A, \mathcal{B}, \preceq)$, once the S-elements $S(f_i, f_j)$ are well-constructed and form a *finite set*, or an "effective" finite set may be "selected" (as in [Mor2], see CH.III §1), an analogue of the Buchberger's algorithm on A can be (at least theoretically first) reached. And as we have seen from the commutative Buchberger's algorithm, if A is not (left) Noetherian, the **While** loop in the noncommutative algorithm may not stop in a finite steps and may consequently yield an infinite Gröbner basis. For instance, if we work on the free algebra $A = k\langle X, Y \rangle$ and set $X >_{grlex} Y$, then, starting with the element $g_0 = X^2 - XY$ yields $g_1 = XYX - XY^2 = S(g_0, g_0)$ and both the programme of [FG] and the algorithm of [Mor2] produce the Gröbner basis $\mathcal{G} = \{g_n = XY^nX - XY^{n+1}\}_{n \geq 0}$ (see §4).

We finish this section by a couple of examples.

Example (i) Set $X >_{grlex} Y$ in the free algebra $k\langle X, Y \rangle$ and consider $\mathcal{G} = \{g_1, g_2\} \subset k\langle X, Y \rangle$ where

$$g_1 = X^2Y - \alpha XYX - \beta YX^2 - \gamma X$$
$$g_2 = XY^2 - \alpha YXY - \beta Y^2X - \gamma Y, \qquad \alpha, \beta, \gamma \in k.$$

Then since $\mathbf{LM}(g_1) = X^2Y$, $\mathbf{LM}(g_2) = XY^2$, and $X \cdot \mathbf{LM}(g_2) \cdot 1 = 1 \cdot \mathbf{LM}(g_1)Y$, it follows that

$$
\begin{aligned}
S(g_2, g_1) &= X \cdot g_2 \cdot 1 - 1 \cdot g_1 \cdot Y \\
&= \beta(YX^2Y - XY^2X) \\
&= 1 \cdot g_2 \cdot X - Y \cdot g_1 \cdot 1.
\end{aligned}
$$

It is easy to see that $\overline{S(g_2, g_1)}^{\mathcal{G}} = 0$. Hence $\mathcal{G}$, which is the set of defining relations of the down-up algebra $A(\alpha, \beta, \gamma)$ (CH.I §5, Example (ix)), is a Gröbner basis in $k\langle X, Y \rangle$. By Theorem 3.1, $A(\alpha, \beta, \gamma)$ has a k-basis $\{u^i(du)^j d^k \mid i, j, k \geq 0\}$.

(ii) Let $>$ be a well-ordering on the set Λ. Set in the free algebra $k\langle X \rangle$ with $X = \{X_i\}_{i \in \Lambda}$ the following

$$R_i = X_i^2, \quad R_{k\ell} = X_kX_\ell + X_\ell X_k, \ k > \ell,$$
$$\mathcal{G} = \{R_i, \ R_{k\ell}\}.$$

If we consider the admissible system $(k\langle X\rangle, \mathcal{B}, \geq_{grlex})$ as defined in §1 Example (iv), where $X_k >_{grlex} X_\ell$ if $k > \ell$, then since $\mathbf{LM}(R_i) = X_i^2$, $\mathbf{LM}(R_{k\ell}) = X_k X_\ell$, it is not hard to see that every S-element obtained by considering different overlap has a Gröbner representation by $\mathcal{G}$, for instance

$$\text{for } k \neq i \neq \ell, \ k > \ell,$$
$$X_i^2 X_k X_\ell - X_i^2 (X_k X_\ell + X_\ell X_k) = -X_i^2 X_\ell X_k;$$
$$\text{for } k > i,$$
$$X_k X_i^2 - (X_k X_i + X_i X_k) X_i = -X_i(X_k X_i + X_i X_k) + X_i^2 X_k.$$

Hence $\mathcal{G}$, which is the set of defining relations of the Grassmann algebra $G(\mathbf{x})$ (CH.I §5, Example (viii)), is a Gröbner basis in $k\langle X\rangle$. By Theorem 3.1, $G(\mathbf{x})$ has a k-basis $\{x_{i_1} x_{i_2} \cdots x_{i_m} \mid i_1 < i_2 < \cdots < i_m, \ i_j \in \Lambda\}$

6. (Left) Dickson Systems and (Left) G-Noetherian Algebras

Let $(A, \mathcal{B}, \preceq)$ be a (left) admissible system in the sense of §1. In this section we characterize the G-Noetherian property of A in the sense: A is G-*Noetherian* if every two-sided ideal of A has a finite Gröbner basis; and A is *left* G-*Noetherian* if every left ideal has a finite Gröbner basis. As a result, we formulate the (left) Dickson systems $(A, \mathcal{B} \preceq)$ in which every (left) ideal has a finite (left) Gröbner basis.

Given a (left) admissible system $(A, \mathcal{B}, \preceq)$, the divisibility (Definition 2.2) yields another order $\preceq'$ on $\mathcal{B}$ as follows:

If $w, u \in \mathcal{B}$ and $w \neq u$, then $w \prec' u$ if and only if $u = \mathbf{LM}(vws)$ for some $v, s \in \mathcal{B}$ with $v \neq 1$ or $s \neq 1$. (For a left admissible system, if and only if $u = \mathbf{LM}(vw)$ for some $v \in \mathcal{B}$ with $v \neq 1$).

Note that $w \preceq' u$ actually implies $w = u$ or $w \prec u$ by Definition 1.2 (**MO2**) (but the converse is not necessarily true). Thus, since $\preceq$ is a well-ordering, we have the following fact.

6.1. Lemma If S is any nonempty subset of $\mathcal{B}$, then the set of minimal elements in S with respect to $\preceq'$, denoted $\min\{S, \preceq'\}$, is nonempty.

$\square$

To characterize the existence of finite Gröbner bases (also see the remark given in the end of this section), we assume from now on that

$(*)$ $\preceq'$ is a *partial ordering* on $\mathcal{B}$, i.e., it is reflexive, transitive and antisymmetric.

Maintaining the notation of §2, if T is a subset of A, we put

$$\mathbf{LM}(T) = \Big\{ \mathbf{LM}(f) \,\Big|\, f \in T \Big\} \subset \mathcal{B},$$

$$\mathbf{LM}^d(T) = \Big\{ \mathbf{LM}^d(f) \,\Big|\, f \in T \text{ not a quasi-zero element} \Big\} \subset \mathcal{B},$$

$$\mathbf{LM}^{\ell d}(T) = \Big\{ \mathbf{LM}^{\ell d}(f) \,\Big|\, f \in T \text{ not a left quasi-zero element} \Big\} \subset \mathcal{B}.$$

It is clear that *if $\mathcal{B}$ is closed under multiplication* or *if A is a domain* then $\mathbf{LM}(T) = \mathbf{LM}^d(T) = \mathbf{LM}^{\ell d}(T)$.

6.2. Theorem Let $\mathcal{F} = \{f_1, ..., f_s\}$ be an s-tuple of nonzero elements in a two-sided ideal I, respectively in a left ideal L of A. With notation and the assumption as above, the following are equivalent.
(i) $\mathcal{F}$ is a Gröbner basis of I, respectively a left Gröbner basis of L;
(ii) $\min\{\mathbf{LM}(I) \cup \mathbf{LM}^d(\mathcal{F}), \preceq'\} = \min\{\mathbf{LM}(\mathcal{F}) \cup \mathbf{LM}^d(\mathcal{F}), \preceq'\}$, respectively $\min\{\mathbf{LM}(L) \cup \mathbf{LM}^{\ell d}(\mathcal{F}), \preceq'\} = \min\{\mathbf{LM}(\mathcal{F}) \cup \mathbf{LM}^{\ell d}(\mathcal{F}), \preceq'\}$.

Proof We only need to prove the equivalence for I because a similar argumentation works for L. To be convenient, we put

$$\min\{\mathbf{LM}(I) \cup \mathbf{LM}^d(\mathcal{F}), \preceq'\} = U, \quad \min\{\mathbf{LM}(\mathcal{F}) \cup \mathbf{LM}^d(\mathcal{F}), \preceq'\} = V.$$

(i) $\Rightarrow$ (ii). Suppose that $\mathcal{F} = \{f_1, ..., f_s\}$ is a Gröbner basis for I. We may assume that each f_i is not a quasi-zero element (see the remark under Proposition 3.2 for the reason).
For $u \in U$, if $u = \mathbf{LM}^d(f_i)$ for some $f_i \in \mathcal{F}$, then clearly $u \in V$. If $u = \mathbf{LM}(g)$ for some $g \in I$, $g \notin \mathcal{F}$, consider a Gröbner representation of g with respect to $\mathcal{F}$:

$$g = \sum_{i=1}^{n} \lambda_i u_i f_i v_i \text{ with } \mathbf{LM}(u_i f_i v_i) \preceq \mathbf{LM}(g) \text{ whenever } u_i f_i v_i \neq 0.$$

By Lemma 1.3, we have $\mathbf{LM}(g) = \mathbf{LM}(u_j f_j v_j)$ for some $j (1 \le j \le n)$, and it follows from Lemma 2.3 that $\mathbf{LM}(g) = \mathbf{LM}(u_j \mathbf{LM}^d(f_j) v_j)$. Thus, $\mathbf{LM}^d(f_j) \preceq' \mathbf{LM}(g) = u$ and hence $u \in V$. This proves $U \subset V$. Conversely, let $v \in V$. Suppose that $g \in I$ is such that $\mathbf{LM}(g) \prec' v$. Then clearly $g \notin \mathcal{F}$. Consider a Gröbner representation of g with respect to $\mathcal{F}$:

$$g = \sum_{i=1}^{n} \lambda_i u_i f_i v_i \text{ with } \mathbf{LM}(u_i f_i v_i) \preceq \mathbf{LM}(g) \text{ whenever } u_i f_i v_i \neq 0.$$

By Lemma 1.3, we have $\mathbf{LM}(g) = \mathbf{LM}(u_j f_j v_j)$ for some $j (1 \le j \le n)$, and it follows from Lemma 2.3 that $\mathbf{LM}(g) = \mathbf{LM}(u_j \mathbf{LM}^d(f_j) v_j)$. Thus, $\mathbf{LM}^d(f_j) \preceq'$

$\mathbf{LM}(g) \prec' v$. By the assumption $(*)$ made on $\preceq'$ we get $\mathbf{LM}^d(f_j) \prec' v$, contradicting the minimality of v. This shows that $v \in U$ and hence $V \subset U$. Therefore $U = V$.

(ii) $\Rightarrow$ (i). Let $f \in I$ be nonzero. We prove that f has a Gröbner representation with respect to $\mathcal{F}$. By the division algorithm (§2) we have $f = \sum_{i=1}^{s} v_i f_i s_i + r$, where $\mathbf{LM}(v_i f_i s_i) \preceq \mathbf{LM}(f)$ whenever $v_i f_i s_i \neq 0$, and $r = \overline{f}^{\mathcal{F}}$. If $r \neq 0$, then since $r \in I$ there exists some $w \in \min\{\mathbf{LM}(I) \cup \mathbf{LM}^d(\mathcal{F}), \preceq'\}$ such that $w \preceq'$ $\mathbf{LM}(r)$. But since $U = V$, there is some $f_j \in \mathcal{F}$ such that $w = \mathbf{LM}(f_j)$ or there is some $f_i \in \mathcal{F}$ such that $w = \mathbf{LM}^d(f_i)$. It is clear that the latter case cannot happen because $r = \overline{f}^{\mathcal{F}}$. Thus, in the first case $\mathbf{LM}(f_j) \neq \mathbf{LM}^d(f_j)$. Then $\mathbf{LM}(f_j)$ is a quasi-zero element and consequently it cannot divide any monomial in $\mathcal{B}$ except for itself, i.e., $w \preceq' \mathbf{LM}(r)$ only when $w = \mathbf{LM}(f_j) = \mathbf{LM}(r)$. But $\mathbf{LM}(f_j) = \mathbf{LM}(r)$ can never happen since $r = \overline{f}^{\mathcal{F}}$. This proves $r = 0$, as desired. $\square$

From the above theorem we retain that if $\mathcal{F} = \{f_1, ..., f_s\}$ is a left Gröbner basis of the left ideal $L = \langle f_1, ..., f_s]$, respectively a Gröbner basis of the ideal $I = \langle f_1, ..., f_s \rangle$, then $\min\{\mathbf{LM}(L) \cup \mathbf{LM}^{\ell d}(\mathcal{F}), \preceq'\}$, respectively $\min\{\mathbf{LM}(I) \cup \mathbf{LM}^d(\mathcal{F}), \preceq'\}$, is a *finite* set. One may hope that the converse of this consequence is also true. Indeed, we have the following more flexible result.

6.3. Proposition Let I, respectively L, be as above. With notation as in Theorem 6.2, if one of the following conditions is satisfied:
(i) $\min\{\mathbf{LM}(I), \preceq'\}$, respectively $\min\{\mathbf{LM}(L), \preceq'\}$, is finite;
(ii) $\min\{\mathbf{LM}(I) \cup \mathbf{LM}^d(\mathcal{F}), \preceq'\}$, respectively $\min\{\mathbf{LM}(L) \cup \mathbf{LM}^{\ell d}(\mathcal{F}) \preceq'\}$, is finite,
then I has a finite Gröbner basis, respectively L has a finite left Gröbner basis in A.

Proof Suppose that the condition (i) is satisfied. (If (ii) is satisfied, the proof is similar.) We may assume that the finitely many minimal elements in $\min\{\mathbf{LM}(I), \preceq'\}$ are $w_1, ..., w_n$. For each w_i, we may choose $f_{w_i} \in I$ with $w_i = \mathbf{LM}(f_{w_i})$. Put

$$\mathcal{F} = \left\{ f_{w_i} \mid i = 1, ..., n \right\}.$$

We now prove that

$$U = \min\{\mathbf{LM}(I) \cup \mathbf{LM}^d(\mathcal{F}), \preceq'\} = \min\{\mathbf{LM}(\mathcal{F}) \cup \mathbf{LM}^d(\mathcal{F}), \preceq'\} = V.$$

If $u \in U$ and $u = \mathbf{LM}(g)$ for some $g \in I$, then there is some $w_i = \mathbf{LM}(f_{w_i})$ such that $w_i \preceq' \mathbf{LM}(g) = u$. Since $f_{w_i} \in \mathcal{F}$, it follows that $u = w_i \in V$. If $u = \mathbf{LM}^d(f_{w_j}) = w_j$ for some $f_{w_j} \in \mathcal{F}$, then clearly $u \in V$. This proves $U \subset V$. Conversely, suppose $v \in V$ and $g \in I$ such that $\mathbf{LM}(g) \prec' v$. Then there exists some $w_i = \mathbf{LM}(f_{w_i})$ such that $w_i \preceq' \mathbf{LM}(g) \prec' v$. By the assumption $(*)$ made on $\preceq'$ this is a contradiction. Thus $v \in U$ and hence $V \subset U$. Therefore, $U = V$. It follows from Theorem 6.2 that $\mathcal{F}$ is a Gröbner basis of I.

A similar argumentation for L finishes the proof.   □

6.4. Definition Suppose that $\leq$ is a quasi-ordering on a set M (i.e., $\leq$ is reflexive and transitive). $\leq$ is called a *Dickson quasi-ordering* if every subset N of M has a finite Dickson basis with respect to $\leq$, that is, there exists a finite subset $S \subset N$ such that for every $u \in N$ there is some $s \in S$ with $s \leq u$.

We refer to ([BW] Ch.4) for details concerning Dickson quasi-orderings. The property of a Dickson partial ordering mentioned in the following proposition is fundamental for the applications of the foregoing results.

6.5. Proposition ([BW] Ch.4 Corollary 4.43) Let $\leq$ be a Dickson partial ordering on M. Then every nonempty subset N of M has a *unique minimal finite basis* S, i.e., a finite basis S such that $S \subseteq C$ for all other bases C of N. S consists of all minimal elements of N.

                                                                                      □

6.6. Theorem Let $(A, \mathcal{B}, \preceq)$ be a (left) admissible system and $\preceq'$ the ordering on $\mathcal{B}$ induced by the divisibility in A. If $\preceq'$ is a Dickson partial ordering on $\mathcal{B}$, then every ideal, respectively every left ideal of A, has a finite Gröbner basis, respectively a finite left Gröbner basis. Consequently, A is G-Noetherian and left G-Noetherian.

Proof This follows from Proposition 6.5 and Proposition 6.3 immediately.   □

6.7. Definition Let $(A, \mathcal{B}, \preceq)$ be a (left) admissible system such that the ordering $\preceq'$ on $\mathcal{B}$ induced by the divisibility in A is a Dickson partial ordering. Then we call $(A, \mathcal{B}, \preceq)$ a (left) *Dickson system*.

Remark So far our discussion has been based on Definition 1.2 in the case that 1 is contained in the k-basis $\mathcal{B}$ of A. If one may pay a little more attention to each of the previous sections, however, it is not hard to see that everything we have done may be modified into the case where $1 \notin \mathcal{B}$ (i.e., by simply modifying a condition used on $\mathcal{B}$ in [Gr]: If $w \in \mathcal{B}$ and w is not a quasi-zero element, then there are $u, v \in \mathcal{B}$ such that $w = \mathbf{LM}^d(uwv)$, so that the division by non-quasi-zero elements is *reflexive*). The reason that we specify this point here is that in practice one does need to construct Gröbner bases by using a k-basis $\mathcal{B}$ not containing 1, for instance, in the path algebras that are widely used in representation theory. To be convenient we now give the definition of a path algebra as follows (e.g., see [Gr]).

Let Γ be a finite   directed graph with $\Gamma_0 = \{v_1, ..., v_s\}$ = vertices and $\Gamma_1 =$

$\{a_1, ..., a_m\}$ = arrows. We let $\mathcal{B} = \{$ finite directed paths in $\Gamma\}$. Then $\Gamma_0 \subset \mathcal{B}$ are viewed as paths of length 0.

The *path algebra*, $A = k\Gamma$, is the k-vector space with basis $\mathcal{B}$. Note that usually $1 \notin \mathcal{B}$. Multiplication on A is defined as: If $p, q \in \mathcal{B}$, then

$$p \cdot q = \begin{cases} 0 \text{ if terminus of } p \neq \text{ origin of } q, \\ pq \in \mathcal{B} \text{ if terminus of } p = \text{ origin of } q. \end{cases}$$

That is, multiplication of paths is given by concatenation, when this makes sense. The length-lexicographic order on $\mathcal{B}$, denoted $>_{lglex}$, is defined as: First fix an arbitrary order on the vertices and arrows, say, $v_s >_{lglex} \cdots >_{lglex} v_1$ and $a_m >_{lglex} \cdots >_{lglex} a_1$. If $p, q \in \mathcal{B}$, then $q >_{lglex} p$ if either the length of $q >$ the length of p or $q = a_{j_1} \cdots a_{j_d}$ and $p = a_{i_1} \cdots a_{i_d}$ then there is an ℓ, $1 \leq \ell \leq d$ such that

$$a_{j_1} = a_{i_1}, ..., a_{j_{\ell-1}} = a_{i_{\ell-1}} \text{ but } a_{j_\ell} >_{lglex} a_{i_\ell}.$$

It is not difficult to verify that $(A, \mathcal{B}, >_{lglex})$ is an admissible system in the sense of §1.

7. Solvable Polynomial Algebras

In this section, we introduce the class of solvable polynomial algebras in the sense of [K-RW], which provides all (left) Dickson systems we will mostly work with in later chapters, and from which we also construct Dickson systems of some algebras with divisors of zero.

First we need some preliminaries.

Let $(M_i, \preceq)$ be quasi-ordered sets, $i = 1, ..., n$, and $M = M_1 \times \cdots \times M_n$ the Cartesian product of M_i. The *direct product* of $(M_i, \preceq)$ is the quasi-ordered set $(M, \preceq')$, where $\preceq'$ is defined by

$$(a_1, ..., a_n) \preceq' (b_1, ..., b_n) \Leftrightarrow a_i \preceq b_i, \ i = 1, ..., n.$$

7.1. Proposition ([BW] CH.4, Corollary 4.47) Let $(M_i, \preceq)$ be Dickson quasi-ordered sets for $i = 1, ..., n$, and let $(M, \preceq')$ be the direct product of the $(M_i, \preceq)$. Then $(M, \preceq')$ is a Dickson quasi-ordered set.

$\square$

A special case of the above proposition is the well-known Dickson's lemma which is mentioned as follows.

Dickson's Lemma Let $(I\!N^n, \preceq')$ be the direct product of n copies of the natural numbers $(I\!N, \prec)$ with their natural ordering. Then, $(I\!N^n, \preceq')$ is a Dickson partially ordered set. More explicitly, every subset S of $I\!N^n$ has a finite subset B such that for every $(m_1, ..., m_n) \in S$, there exists $(k_1, ..., k_n) \in B$ with $k_i \leq m_i$ for $i = 1, ..., n$.

$\square$

Let $A = k[a_1, ..., a_n]$ be a finitely generated k-algebra with n generators $a_1, ..., a_n$, and let

$$B = \left\{ a^\alpha = a_1^{\alpha_1} \cdots a_n^{\alpha_n} \mid \alpha = (\alpha_1, ..., \alpha_n) \in I\!N^n \right\}$$

be the set of all standard monomials in A (see §1).

7.2. Proposition Let A be as above and satisfy

(1) B is a k-basis of A (then $1 \in B$), and
(2) $a^\alpha a^\beta = \lambda_{\alpha,\beta} a^{\alpha+\beta} + f$, $a^\alpha, a^\beta \in B$, $\lambda_{\alpha,\beta} \in k - \{0\}$, $f \in A$.

The following holds.
(i) If $(A, B, \preceq)$ is an admissible system such that in the above (2)

$$\mathbf{LM}(f) \prec a^{\alpha+\beta},$$

then $(A, B, \preceq)$ is a Dickson system and a left Dickson system as well.
(ii) If $(A, B, \preceq)$ is a left admissible system such that in the above (2)

$$\mathbf{LM}(f) \prec a^{\alpha+\beta},$$

then $(A, B, \preceq)$ is a left Dickson system; If furthermore $\preceq$ is either $\geq_{lex}$ or a graded monomial ordering $\succeq_{gr}$ (see §1), then $(A, B, \preceq)$ is also an admissible system and hence a Dickson system.

Proof Let $\preceq'$ be the (partial) ordering on B induced by the divisibility in A (§6 (∗)). By the assumption we have, for $a^\beta, a^\alpha \in B$,

a^β is divisible by a^α, i.e., $a^\alpha \preceq' a^\beta$, if and only if
$a^\beta = \mathbf{LM}(a^\gamma a^\alpha a^\eta) = a^{\gamma+\alpha+\eta}$ if and only if
$\beta = \alpha + \delta$ for some $\delta \in I\!N^n$.

It follows from Dickson's lemma that $\preceq'$ is a Dickson partial ordering on B. Hence $(A, B, \preceq)$ is a Dickson system. Since $(A, B, \preceq)$ is also a left admissible system (see §1), a similar argumentation as above shows that $(A, B, \preceq)$ is also a left Dickson system. But then (ii) follows as well. $\square$

Let $A = k[x_1, ..., x_n]$ be the commutative polynomial k-algebra in n variables, and $B = \{x_1^{\alpha_1} \cdots x_n^{\alpha_n} \mid (\alpha_1, ..., \alpha_n) \in I\!N^n\}$ the standard k-basis for A. Then any admissible system $(A, B, \preceq)$ is a Dickson system. (Note that in this case a left Dickson system is the same as a Dickson system.)

7.3. Definition (A. Kandri-Rody and V. Weispfenning, 1990) Let A be as in Proposition 7.2. We call A a *solvable polynomial algebra*, provided A satisfies (i) or (ii) of Proposition 7.2.

From Proposition 7.2 we immediately derive the following fact.

7.4. Proposition (i) Solvable polynomial algebras are (left) Noetherian domains.
(ii) For any nonzero f and g in a solvable polynomial algebra A with $\mathbf{LM}(f) = a^\alpha = a_1^{\alpha_1} \cdots a_n^{\alpha_n}$, $\mathbf{LM}(g) = a^\beta = a_1^{\beta_1} \cdots a_n^{\beta_n}$, putting

$$\gamma_i = \max\{\alpha_i, \beta_i\}_{i=1}^n, \quad a^\gamma = a_1^{\gamma_1} \cdots a_n^{\gamma_n},$$
$$\lambda = \mathbf{LC}(a^{\gamma-\alpha}f), \quad \mu = \mathbf{LC}(a^{\gamma-\beta}g),$$

then a *unique* left S-element of f and g is defined as

$$S(f,g) = \frac{\mu}{\lambda}a^{\gamma-\alpha}f - a^{\gamma-\beta}g.$$

□

The class of solvable polynomial algebras includes CH.I §5 Example (i)–(vii) as well as many other skew polynomial algebras (§1 Example (ii), also see CH.III §2). In particular, with respect to *any* monomial ordering on $I\!N^n$, a multiplicative analogue $\mathcal{O}_n(\lambda_{ji})$ of the Weyl algebra (CH.I §5 Example (iii) is a solvable polynomial algebra.
If A is a solvable polynomial algebra with respect to some *graded* monomial ordering $\succeq_{gr}$, then $(A, \mathcal{B}, \succeq_{gr})$ is a Dickson system, and hence A is both G-Noetherian and left G-Noetherian.
More examples of solvable polynomial algebras are given in next chapter.

We now proceed to construct examples of (left) Dickson systems in which the algebras contain divisors of zero.

Example (i) Consider the Grassmann algebra $G(\mathbf{x})$ as defined in CH.I §5 with $\mathbf{x} = \{x_i\}_{i \in \Lambda}$. Then there exists a (left) admissible system $(G(\mathbf{x}), \mathcal{B}, \preceq)$; if $|\Lambda| = n$ is finite, then $(G(\mathbf{x}), \mathcal{B}, \preceq)$ is a (left) Dickson system.

Proof From §5 Example (ii) we know that the set of defining relations of $G(\mathbf{x})$, denoted $\mathcal{G} = \{R_i, R_{k\ell}\}$, is a (left) Gröbner basis in $k\langle X \rangle$ and that $\mathcal{B} = \{x_{i_1} x_{i_2} \cdots x_{i_m} \mid i_j \in \Lambda, i_1 < i_2 < \cdots < i_m\}$ forms a k-basis for $G(\mathbf{x})$. It is easy to see that the monomial ordering $\geq_{grlex}$ in $k\langle X \rangle$ induces a graded lex ordering on $\mathcal{B}$. Hence $(G(\mathbf{x}), \mathcal{B}, \geq_{grlex})$ forms a (left) admissible system. In the case where $|\Lambda| = n$, we see that

$$\mathcal{B} = \left\{ x_1^{\alpha_1} x_2^{\alpha_2} \cdots x_n^{\alpha_n} \mid \alpha_i = 0, 1 \right\}$$

and $|\mathcal{B}| = 1 + \begin{pmatrix} n \\ 1 \end{pmatrix} + \begin{pmatrix} n \\ 2 \end{pmatrix} + \cdots + \begin{pmatrix} n \\ n-1 \end{pmatrix} + 1 = 2^n = \dim_k G(\mathbf{x})$.
Observing the multiplication (and hence division) in $G(\mathbf{x})$, $(G(\mathbf{x}), \mathcal{B}, \geq_{grlex})$ is clearly a (left) Dickson system. □

As we pointed out in CH.I §5 (or from the above proof), in the case where $\mathbf{x}$ is a finite set, $G(\mathbf{x})$ is indeed a homomorphic image of a multiplicative analogue $\mathcal{O}_n(\lambda_{ji})$ of the Weyl algebra.
More generally, let $R = k[x_1,...,x_n,x_{n+1},...,x_{n+m}]$ be a solvable polynomial k-algebra and $(R, \mathcal{B}, \preceq)$ an admissible system associated with R such that

$$x_j x_i = \lambda_{ij} x_i x_j, \ 1 \leq i < j \leq n+m, \ \lambda_{ij} \in k - \{0\}.$$

Then by Proposition 4.1, $\mathcal{G} = \{x_{n+1}^p, x_{n+2}^p, ..., x_{n+m}^p\}$ forms a Gröbner basis for the two-sided ideal $I = \langle x_{n+1}^p, ..., x_{n+m}^p \rangle$, where p is a fixed positive integer (also see CH.V §1 Propositon 1.4). Consider the quotient algebra $A = R/I$ and write a_i for the coset of each x_i in A, $1 \leq i \leq n$, and write b_j for the coset of each x_{n+j} in A, $1 \leq j \leq m$. Since $\mathcal{B} = \{x_1^{\alpha_1} \cdots x_n^{\alpha_n} x_{n+1}^{\beta_1} \cdots x_{n+m}^{\beta_m} \mid \alpha = (\alpha_1,...,\alpha_n) \in \mathbb{N}^n, \beta = (\beta_1,...,\beta_m) \in \mathbb{N}^m\}$ and $\mathcal{G}$ is a Gröbner basis for I, it follows from Theorem 3.1 and Proposition 3.3 that

$$\overline{\mathcal{B}} = \left\{ a_1^{\alpha_1} \cdots a_n^{\alpha_n} b_1^{\beta_1} \cdots b_m^{\beta_m} \mid (\alpha_1,...,\alpha_n) \in \mathbb{N}^n, 0 \leq \beta_i \leq p-1 \right\}$$

is a k-basis for A. It is clear that A is no longer a solvable polynomial algebra. But by the assumptions on R we easily derive the following result.

7.5. Proposition $(A, \overline{\mathcal{B}}, \preceq)$ is an admissible system and hence a Dickson system if $\preceq$ denotes the ordering $\geq_{lex}$ or some graded monomial ordering $\succeq_{gr}$ on $\overline{\mathcal{B}}$.

8. No (Left) Monomial Ordering Existing on $\Delta(k[x_1,...,x_n])$ with char$k > 0$

This final section is focused on the algebra of linear partial differential operators with polynomial coefficients because of its importance in practice.

Let k be a field, $k[x_1,...,x_n]$ the commutative polynomial algebra in n variables and $\text{End}_k k[x_1,...,x_n]$ the k-algebra of all k-linear operators on $k[x_1,...,x_n]$. Recall that the algebra of k-linear differential operators with polynomial coefficients, denoted $\Delta(k[x_1,...,x_n])$, is the subalgebra of $\text{End}_k k[x_1,...,x_n]$ generated by $k[x_1,...,x_n]$ and all partial derivations $\partial_1 = \partial/\partial x_1, ..., \partial_n = \partial/\partial x_n$, where each x_i acts on $k[x_1,...,x_n]$ by the left multiplication. More precisely,

$$\Delta(k[x_1,...,x_n]) = k[x_1,...,x_n,\partial_1,...,\partial_n] \subset \text{End}_k k[x_1,...,x_n],$$

and the generators of $\Delta(k[x_1, ..., x_n])$ satisfy

$$\partial_j x_i = x_i \partial_j + \delta_{ij} \text{ (the Kronecker delta)}, \ i, j = 1, ..., n,$$
$$x_i x_j = x_j x_i, \quad \partial_i \partial_j = \partial_j \partial_i, i, j = 1, ..., n.$$

If $\text{char} k = 0$, then it is well-known (e.g., [Bj], [MR]) that

(i) The set of all standard monomials of $\Delta(k[x_1, ..., x_n])$, denoted

$$\mathcal{B} = \left\{ x_1^{\alpha_1} \cdots x_n^{\alpha_n} \partial_1^{\beta_1} \cdots \partial^{\beta^n} \ \middle| \ (\alpha_1, ..., \alpha_n, \beta_1, ..., \beta_n) \in I\!N^{2n} \right\},$$

is a k-basis of $\Delta(k[x_1, ..., x_n])$.

(ii) $\Delta(k[x_1, ..., x_n])$ coincides with the ring of differential operators $\mathcal{D}(\mathbf{A}_k^n)$ of the affine n-space $\mathbf{A}_k^n$ over k.

(iii) (See CH.I §5, Example (i)) $\Delta(k[x_1, ..., x_n])$ is isomorphic to the nth Weyl algebra $A_n(k) = k[x_1, ..., x_n, y_1, ..., y_n]$ generated by $2n$ elements subject to the relations:

$$y_j x_i = x_i y_j + \delta_{ij} \text{ (the Kronecker delta)}, \quad 1 \leq i, j \leq n,$$
$$x_i x_j = x_j x_i, \ y_i y_j = y_j y_i, \qquad\qquad 1 \leq i < j \leq n.$$

(iv) (See CH.I §5, Example (i)) $\Delta(k[x_1, ..., x_n])$ is isomorphic to the iterated skew polynomial k-algebra

$$k[x_1, ..., x_n][y_1; \delta_1] \cdots [y_n; \delta_n],$$

where $\delta_i = \partial/\partial x_i, \ i = 1, ..., n$.

(v) $\Delta(k[x_1, ..., x_n])$ is an infinite dimensional simple k-algebra and is a Noetherian domain.

It follows from §7 that $(\Delta(k[x_1, ..., x_n]), \mathcal{B}, \preceq)$ is a (left) admissible system and hence it is a (left) Dickson system (at least when $\preceq$ is $\geq_{lex}$ or $\geq_{grlex}$), for, in this case, $\Delta(k[x_1, ..., x_n])$ is a solvable polynomial algebra.

However, if $\text{char} k = p > 0$ for some prime number p, then $\Delta(k[x_1, ..., x_n])$ no longer has the nice properties 1–4 listed above, in particular, it is not a solvable polynomial k-algebra in the sense of §7. This follows from the following facts. (The following 8.1–8.3, and 8.5 are obtained in [Zh].)

It is easy to see that every element $f \in \Delta(k[x_1, ..., x_n])$ is of the form

$$f = \sum_{\alpha, \beta} \lambda_{\alpha\beta} x_1^{\alpha_1} \cdots x_n^{\alpha_n} \partial_1^{\beta_1} \cdots \partial_n^{\beta_n}$$

where $\lambda_{\alpha\beta} \in k, \ \alpha = (\alpha_1, ..., \alpha_n), \ \beta = (\beta_1, ..., \beta_n) \in I\!N^n$.

As in §1 we write $|\alpha| = \alpha_1 + \cdots + \alpha_n, \ |\beta| = \beta_1 + \cdots + \beta_n, \ x^\alpha = x_1^{\alpha_1} \cdots x_n^{\alpha_n}, \ \partial^\beta = \partial_1^{\beta_1} \cdots \partial_n^{\beta_n}$. Moreover, we write $\alpha! = \alpha_1! \cdots \alpha_n!$, $0! = 1$, and write $D * f$ for the action of an operator D on an element of $f \in k[x_1, ..., x_n]$.

8.1. Lemma With the notation as above, the following holds.
(i) $\partial_i^p = 0$, $i = 1, ..., n$.
(ii) For any $\alpha, \beta \in I\!N^n$, if $|\beta| \geq |\alpha|$, then

$$\partial^\beta * x^\alpha = \begin{cases} \beta! & (\beta = \alpha \text{ and } \beta_i \leq p - 1, \ 1 \leq i \leq n) \\ 0 & \text{otherwise.} \end{cases}$$

$\square$

For the reader's convenience we include a proof of the next proposition.

8.2. Proposition With the notation as above,

$$\mathcal{B} = \left\{ x^\alpha \partial^\beta \ \middle| \ \alpha = (\alpha_1, ..., \alpha_n), \beta = (\beta_1, ..., \beta_n) \in I\!N^n, \ 0 \leq \beta_i \leq p - 1 \right\}$$

forms a k-basis for $\Delta(k[x_1, ..., x_n])$.

Proof Suppose that $D = \sum_{(\alpha,\beta)} \lambda_{(\alpha,\beta)} x^\alpha \partial^\beta = 0$, where the (α, β) are mutually different and $\lambda_{(\alpha,\beta)} \in k$. Rewrite D as a polynomial in ∂ with coefficients in $k[x_1, ..., x_n]$: $D = \sum_{i=1}^{t} f_i \partial^{\gamma^{(i)}}$, where $f_i = \sum_{j=1}^{\ell} \lambda_{\alpha^{(j)}\gamma^{(i)}} x^{\alpha^{(j)}}$, $i = 1, ..., t$. Note that in the expression of f_i the $x^{\alpha^{(j)}}$ are mutually different. Using the graded lexicographic ordering on $I\!N^n$, we may assume that $\gamma^{(1)} >_{grlex} \gamma^{(2)} \ _{grlex} \cdots >_{grlex} \gamma^{(t)}$. Then $0 = D * x^{\gamma^{(1)}} = f_1 \gamma^{(1)}!$. Since $\gamma_i^{(1)} \leq p - 1$, $i = 1, ..., n$, we have $f_1 = 0$. Hence $\lambda_{\alpha^{(j)}\gamma^{(i)}} = 0$, $j = 1, ..., \ell$. Similarly we may get $f_2 = \cdots = f_t = 0$. Therefore, $\lambda_{\alpha,\beta} = 0$ for all (α, β) appearing in the expression of D. This proves the linear independence of elements in $\mathcal{B}$, as desired. $\square$

It follows from Lemma 8.1, Proposition 8.2 and its proof that we also have the following.

8.3. Corollary (i) The centre of $\Delta(k[x_1, ..., x_n])$ is the subalgebra $k[x_1^p, ..., x_n^p]$.
(ii) $\Delta(k[x_1, ..., x_n])$ is not a simple algebra.
(iii) $\Delta(k[x_1, ..., x_n])$ is a free module of finite rank over its centre.

$\square$

It is not hard to see that if there was a (left) monomial ordering $\preceq$ on $\mathcal{B}$, then $(\Delta(k[x_1, ..., x_n]), \mathcal{B}, \preceq)$ would be a (left) Dickson system. Unfortunately, we have the following bad news (from an online discussion of the author with Zhang Jiangfeng in the beginning of Nov. 2000).

8.4. Conclusion Let $\Delta(k[x_1, ..., x_n])$ be as above, where k is of characeristic $p > 0$. Then there does not exist a monomial ordering or a left monomial ordering on

$$\mathcal{B} = \left\{ x_1^{\alpha_1} \cdots x_n^{\alpha_n} \partial_1^{\beta_1} \cdots \partial_n^{\beta_n} \ \middle| \ (\alpha_1, ..., \alpha_n) \in I\!N^n, \ 0 \le \beta_i \le p - 1 \right\}.$$

Proof Consider $u = x_i \partial_i^{p-1}, w = \partial_i^{p-2} \in \mathcal{B}$. If there was a left monomial ordering $\preceq$ on $\mathcal{B}$, then since $\partial_i x_i = x_i \partial_i + 1$ and $\partial_i^p = 0$, the formulas

$$\partial_i u = \partial_i^{p-1}, \qquad \partial_i w = \partial_i^{p-1}$$

would yield a contradiction, i.e., $\partial_i^{p-1} \prec \partial_i^{p-1}$.

Since any monomial ordering on $\mathcal{B}$ is also a left monomial ordering, the above stone has killed two birds. $\qquad\square$

Nevertheless, light still comes, that is, we have two ways to use Gröbner bases on $\Delta(k[x_1, ..., x_n])$ where $\mathrm{char}k = p > 0$. To see this, put $A = \Delta(k[x_1, ..., x_n])$, and consider the standard filtration FA on A which is by definition the filtration: $F_0 A \subset F_1 A \subset \cdots \subset F_n A \subset \cdots$ with

$$F_n A = \left\{ f = \sum_{\alpha,\beta} \lambda_{\alpha\beta} x_1^{\alpha_1} \cdots x_n^{\alpha_n} \partial_1^{\beta_1} \cdots \partial_n^{\beta_n} \ \middle| \ |\alpha| + |\beta| \le n \right\}, \quad n \ge 0.$$

The associated graded algebra of A, denoted $G(A)$, which is by definition the positively graded algebra $G(A) = \oplus_{n \in I\!N} G(A)_n$ with $G(A)_n = F_n A / F_{n-1} A$.

8.5. Proposition (i) A is a homomorphic image of the iterated skew polynomial algebra (i.e., the nth Weyl algebra over k)

$$A_n(k) = k[x_1, ..., x_n][y_1; \delta_1] \cdots [y_n; \delta_n],$$

where $\delta_i = \partial/\partial x_i$, $i = 1, ..., n$. More precisely, $A \cong A_n(k)/I$ where $I = \langle \delta_1^p, ..., \delta_n^p \rangle$.

(ii) $G(A) \cong k[z_1, ..., z_{2n}]/J$, where $k[z_1, ..., z_{2n}]$ is the commutative polynomial k-algebra in $2n$ variables and $J = \langle z_{n+1}^p, ..., z_{2n}^p \rangle$. $\qquad\square$

Note that $A_n(k)$ is a solvable polynomial algebra. From Proposition 8.5(i) it is therefore clear that we can apply Gröbner bases to $\Delta(k[x_1, ..., x_n])$ by passing to $A_n(k)$ (see [Zh] for some applications). On the other hand, from a lifting structure point of view, that is, studying $\Delta(k[x_1, ..., x_n])$ by passing to its associated graded algebra whenever the standard filtration on $\Delta(k[x_1, ..., x_n])$ is considered (see CH.III §3 for a general explanation on this view), Proposition 8.5(ii) now provides us with the possibility to play with the associated graded algebra of $\Delta(k[x_1, ..., x_n])$, namely, it follows from Proposition 7.5 and Proposition 8.5(ii) that we have the following fact.

8.6. **Proposition** Let $A = \Delta(k[x_1, ..., x_n])$ be as before, and consider the standard filtration FA on A. If $\sigma(x_i)$ and $\sigma(\partial_j)$ denote the image of x_i and ∂_j in $G(A)_1 = F_1A/F_0A$ respectively, $i, j = 1, ..., n$, then $(G(A), \mathcal{B}, \preceq)$ is a Dickson system, where

$$\mathcal{B} = \left\{ \sigma(x_1)^{\alpha_1} \cdots \sigma(x_n)^{\alpha_n} \sigma(\partial_1)^{\beta_1} \cdots \sigma(\partial_n)^{\beta_n} \;\middle|\; (\alpha_1, ..., \alpha_n) \in \mathbb{N}^n,\ 0 \leq \beta_i \leq p - 1 \right\}$$

and $\preceq$ denotes the ordering $\geq_{lex}$ or $\geq_{grlex}$ on $\mathcal{B}$.

CHAPTER III
Gröbner Bases and Basic Algebraic-Algorithmic Structures

In view of CH.II §3, we start this chapter with the interaction between the PBW bases of finitely generated algebras and the algorithmic aspect of very noncommutative Gröbner bases (in the sense of [Mor2]). As a consequence, Theorem 1.5 and Proposition 1.6 enable us to recognize and construct quadric solvable polynomial algebras in an algorithmic way. And then, we determine the homogeneous defining relations of the associated graded structures of a given algebra, based on a combination of noncommutative Gröbner basis and the structural trick developed in CH.I §4. The latter result is indeed the noncommutative version of a remarkable application of the commutative Gröbner basis theory to algebraic geometry in determining the homogeneous defining equations of the projective closure of an affine variety. In addition to providing some basic algebraic-algorithmic structures for later study, this chapter also motivates a general filtered-graded transfer of Gröbner bases in associative algebras – the topic of next chapter.

1. PBW Bases of Finitely Generated Algebras

Let k be a field and $A = k[a_1, ..., a_n]$ a finitely generated k-algebra. We say that A *has a* PBW k-*basis* if the set of standard monomials

$$\mathcal{B} = \left\{ a_1^{\alpha_1} a_2^{\alpha_2} \cdots a_n^{\alpha_n} \;\middle|\; (\alpha_1, ..., \alpha_n) \in I\!N^n \right\}$$

forms a k-basis for A. Note that we say "a PBW k-basis" instead of "the PBW k-basis" because the presently used "natural order" $a_1, a_2, ..., a_n$ on generators may be changed by any permutation of generators if it is necessary.

We have seen from CH.II that having a k-basis consisting of "monomials" is the basis of having a Gröbner basis theory for a k-algebra A. It is also well known that if a finitely generated k-algebra A has a PBW k-basis, then the structure theory of A, in particular, the representation theory of A will be more nicer. Inspired by the work of [Mor2] and Berger's quantum PBW theorem [Ber], in this section, we demonstrate the interaction between the algorithmic aspect of very noncommutative Gröbner bases and PBW bases for finitely generated algebras with specific defining relations.

To better understand this section, we suggest the reader to go back to CH.II §3.

First recall from [Mor2] some generalities of the noncommutative Gröbner bases for *two-sided* ideals in free algebras.

Let $k\langle X \rangle$ be the free algebra generated by $X = \{X_i\}_{i \in \Lambda}$ over the field k, and let $S = \langle X \rangle$ be the free semigroup generated by X. With notation as in CH.II, we let $(k\langle X \rangle, \mathcal{B}, \preceq)$ be an *admissible system*, where $\mathcal{B}$ is the k-basis of $k\langle X \rangle$ consisting of words of S and 1. It follows from CH.II Proposition 4.2 that the following definition is necessarily to be mentioned (but it is better to bear CH.II Definition 3.2 in mind as well).

1.1. Definition Let I be an ideal of $k\langle X \rangle$. A set $\mathcal{G} = \{g_j\}_{j \in J} \subset I$ is called a *Gröbner basis* of I if $\langle \mathbf{LM}(\mathcal{G}) \rangle = \langle \mathbf{LM}(I) \rangle$ where $\langle \mathbf{LM}(\mathcal{G}) \rangle$ is the two-sided ideal generated by $\{\mathbf{LM}(g_j) \mid g_j \in \mathcal{G}\}$, and a similar interpretation is for $\langle \mathbf{LM}(I) \rangle$.

Given a generating set $\mathcal{G} = \{g_i\}_{i \in J}$ of an ideal $I \subset k\langle X \rangle$, it is generally difficult to know if $\mathcal{G}$ is a Gröbner basis of I. However, if $k\langle X \rangle$ is the free k-algebra generated by a finite set of indeterminates $X = \{X_1, ..., X_n\}$, and if $\mathcal{G} = \{g_1, ..., g_s\}$ is also finite, then from [Mor2] we know that the noncommutative version of Buchberger's algorithm does exist and it can be used to produce a Gröbner basis of I starting with $\mathcal{G}$, though the obtained basis is usually no longer finite (unless I has a finite Gröbner basis and the procedure halts).   The existence of a   finite

Gröbner basis is based on a technical process in analysing the S-elements (see CH.II §5) that we recall in some detail below.

Let $X = \{X_1, ..., X_n\}$, $(k\langle X\rangle, \mathcal{B}, \preceq)$, and $\mathcal{G} = \{g_1, ..., g_s\}$ be fixed as above but we assume that all g_i are *monic*, i.e.,

$$\mathbf{LC}(g_i) = 1, \; i = 1, ..., s.$$

Adopting the notation as in [Mor2], let $S \times S$ be the Cartesian square of the free semigroup S generated by X. If $(l, r), (\lambda, \rho) \in S \times S$ are such that

$$l\mathbf{LM}(g_j)r = \lambda\mathbf{LM}(g_i)\rho,$$

then an S-element of g_j and g_i is denoted by

$$S(i, j; l, r; \lambda, \rho) = lg_jr - \lambda g_i\rho.$$

We say that $S(i, j; l, r; \lambda, \rho)$ has a *weak Gröbner representation* by $\mathcal{G}$ if

$$S(i, j; l, r; \lambda, \rho) = \sum_{k, \mu} c_{k\mu}l_{k\mu}g_kr_{k\mu} \text{ in which}$$

$$\text{for each } k, \mu, \; l_{ku}\mathbf{LM}(g_k)r_{k\mu} \prec l\mathbf{LM}(g_j)r.$$

The product $S \times S$ has a natural S-bimodule structure in the sense that for each $t \in S$, for each $(l, r) \in S \times S$, $t(l, r) = (tl, r)$, $(l, r)t = (l, rt)$. To be convenient, we denote this algebraic structure on $S \times S$ by $S \otimes S$.

An ideal of $S \otimes S$ is a subset $\mathcal{J} \subset S \otimes S$ such that if $(l, r) \in \mathcal{J}$, $t \in S$, then $(tl, r) \in \mathcal{J}$, $(l, rt) \in \mathcal{J}$; a set of generators for $\mathcal{J}$ is a (not necessarily finite) set $\mathcal{G} \subset \mathcal{J}$ such that for each $(l, r) \in \mathcal{J}$ there are $l_1, r_1 \in S$, $(w_l, w_r) \in \mathcal{G}$ such that $l = l_1w_l$ and $r = w_rr_1$.

For $s \geq j \geq 1$, we write $ST(\mathbf{LM}(g_j))$ for the ideal of $S \otimes S$ generated by the set

$$SOB(\mathbf{LM}(g_j)) = \left\{(1, r) \in S \otimes S \; \middle| \; \begin{array}{l} r \neq 1 \text{ and there is } (l, 1) \in S \otimes S \\ \text{such that } l\mathbf{LM}(g_j) = \mathbf{LM}(g_j)r \end{array}\right\},$$

and we put

$$T_j(\mathcal{G}) = \left\{(l, r) \in S \otimes S \; \middle| \; l\mathbf{LM}(g_j)r \in \mathbf{I}_j\right\} \cup ST(\mathbf{LM}(g_j)),$$

where $\mathbf{I}_j$ stands for the ideal of S generated by $\{\mathbf{LM}(g_1), \mathbf{LM}(g_2), ..., \mathbf{LM}(g_{j-1})\}$. (One may see that $T_j(\mathcal{G})$ is indeed an ideal of $S \otimes S$.)

Let $\mathcal{U}$ be a minimal generating set of the ideal $T_j(\mathcal{G})$. For each $\sigma = (l_\sigma, r_\sigma) \in \mathcal{U}$, choose $i_\sigma, \lambda_\sigma, \rho_\sigma$ such that

$$\begin{aligned} l_\sigma\mathbf{LM}(g_j)r_\sigma &= \lambda_\sigma\mathbf{LM}(g_{i_\sigma})\rho_\sigma, \; i_\sigma < j \text{ or } i_\sigma = j \text{ and} \\ &\qquad \text{there is } w \in S \text{ such that } r_\sigma = w\rho_\sigma, \end{aligned}$$

and we let

$$\text{MIN}(j) = \{(i_\sigma, j; l_\sigma, r_\sigma; \lambda_\sigma, \rho_\sigma)\}.$$

An element $(i, j; l, r; \lambda, \rho) \in \text{MIN}(j)$ is said to be *trivial* if there is $w \in S$ such that either $l = \lambda\mathbf{LM}(g_i)w$ (and so $\rho = w\mathbf{LM}(g_j)r$) or $\lambda = l\mathbf{LM}(g_j)w$ (and so $r = w\mathbf{LM}(g_i)\rho$).

1.2. Theorem ([Mor2] Corollary 5.8, Theorem 5.9) Let $\mathcal{G} = \{g_1, ..., g_s\}$ be a generating set of the ideal $I \subset k\langle X \rangle$ where $\mathbf{LC}(g_i) = 1$ for each i. The following holds.
(i) The set

$$OBS(j) = \{(i, j; l, r; \lambda, \rho) \in \text{MIN}(j) \text{ and nontrivial}\}$$

is finite.
(ii) $\mathcal{G}$ is a Gröbner basis of I if and only if for each j, for each nontrivial $(i, j; l, r; \lambda, \rho) \in \text{MIN}(j)$, the S-element $\text{S}(i, j; l, r; \lambda, \rho)$ has a weak Gröbner representation.

$$\square$$

1.3. Corollary (compare with the example given in the end of CH.II §4) If $A = k\langle X, Y \rangle / I$ with $I = \langle R = YX - \lambda XY - F \rangle$, where $\lambda \in k$, $F \in k\langle X, Y \rangle$, and if $\mathbf{LM}(R) = YX$ with respect to some monomial ordering $\succeq$ on $k\langle X, Y \rangle$, then $\{R\}$ is a Gröbner basis for I and A has the PBW k-basis $\{x^\alpha y^\beta \mid (\alpha, \beta) \in I\!\!N^2\}$, where x and y are the images of X and Y in A respectively.

$$\square$$

Now let $A = k[a_1, ..., a_n]$ be a finitely generated k-algebra and $(k\langle X \rangle, \mathcal{B}, \preceq)$ an admissible system as before, where $k\langle X \rangle = k\langle X_1, ..., X_n \rangle$. Suppose that A is defined by the following relations

$$R_{ji} = X_j X_i - \lambda_{ji} X_i X_j - \{X_j, \ X_i\}, \quad 1 \le i < j \le n,$$

where $\lambda_{ji} \in k$, $\{X_j, \ X_i\} = 0$ or $\{X_j, \ X_i\} \in k\langle X \rangle - k\text{-span}\{X_j X_i, \ X_i X_j\}$. Writing I for the ideal of $k\langle X \rangle$ generated by $\mathcal{G} = \{R_{ji} \mid 1 \le i < j \le n\}$, then $A = k\langle X \rangle / I$. Furthermore, we assume that, with respect to the monomial ordering $\preceq$ in the admissible system $(k\langle X \rangle, \mathcal{B}, \preceq)$, the defining relations satisfy

$$(*) \qquad\qquad \mathbf{LM}(R_{ji}) = X_j X_i, \quad 1 \le i < j \le n.$$

1.4. Lemma With notation as before, if we put

$$\mathcal{G} = \Big\{g_k = R_{kj}, \ g_j = R_{ji} \ \Big| \ 1 \le j, k \le n\Big\},$$

then

$$OBS(k) = \Big\{(h, k; l, r; \lambda, \rho) \in \mathrm{MIN}(k) \text{ and nontrivial}\Big\}$$

$$= \Big\{(j, k; 1, X_i; X_k, 1) \mid i < j < k\Big\}.$$

Proof Recall that

$$T_k(\mathcal{G}) = \Big\{(l, r) \in S \otimes S \mid l\mathbf{LM}(g_k)r \in \mathbf{I}_k\Big\} \cup ST(\mathbf{LM}(g_k)),$$

where $\mathbf{I}_k$ stands for the ideal of S generated by $\{\mathbf{LM}(g_1), \mathbf{LM}(g_2), ..., \mathbf{LM}(g_{k-1})\}$, and $ST(\mathbf{LM}(g_k))$ is the ideal of $S \otimes S$ generated by the set

$$SOB(\mathbf{LM}(g_k)) = \left\{(1, r) \in S \otimes S \ \left|\ \begin{array}{l} r \neq 1 \text{ and there is } (l, 1) \in S \otimes S \\ \text{such that } l\mathbf{LM}(g_k) = \mathbf{LM}(g_k)r \end{array}\right.\right\}.$$

It is easy to see that $SOB(\mathbf{LM}(g_k)) = \{(1, wX_kX_j) \mid w \in S\}$ and $SOB(\mathbf{LM}(g_k))$ is a minimal generating set of $ST(\mathbf{LM}(g_k))$. It is also not hard to check that, for each $(1, wX_kX_j) \in SOB(\mathbf{LM}(g_k))$, every $(j, k; 1, wX_kX_j; \lambda, \rho) \in \mathrm{MIN}(j)$ is trivial. Furthermore if $(l, r) \in S \otimes S$ is such that

$$lX_kX_jr = \lambda X_tX_i\rho \text{ for some } \lambda, \rho \in S \otimes S, \text{ where } k > l,$$

then one sees that $(t, k; l, r; \lambda, \rho)$ is trivial in case $j \neq t$; In the case where $j = t$, one may also easily see that the $(1, X_i)$ generate all nontrivial elements in $\mathrm{MIN}(k)$, or more precisely, $OBS(k) = \{(h, k; l, r; \lambda, \rho) \in \mathrm{MIN}(k) \text{ and nontrivial}\}$ $= \{(j, k; 1, X_i; X_k, 1) \mid i < j < k\}$, as desired. $\square$

1.5. Theorem Let $A = k[a_1, ..., a_n] = k\langle X\rangle/I$ be the finitely generated k-algebra as above, where $I = \langle R_{ji} \mid 1 \leq i < j \leq n\rangle$. Suppose that the R_{ji} satisfy the foregoing assumption $(*)$ with respect to the monomial ordering $\preceq$ in $(k\langle X\rangle, \mathcal{B}, \preceq)$. The following statements are equivalent:
(i) The k-algebra A has a PBW k-basis.
(ii) $\mathcal{G} = \{R_{ji} \mid 1 \leq i < j \leq n\}$ forms a Gröbner basis for the ideal I in $k\langle X\rangle$.
(iii) For $1 \leq i < j < k \leq n$, every $R_{kj}X_i - X_kR_{ji}$ has a weak Gröbner representation by $\mathcal{G}$.

Proof By Lemma 1.4, we have

$$OBS(k) = \{(h, k; l, r; \lambda, \rho) \in \mathrm{MIN}(k) \text{ and nontrivial}\}$$

$$= \Big\{(j, k; 1, X_i; X_k, 1) \mid i < j < k\Big\}.$$

Also note that for each $(j, k; 1, X_i; X_k, 1) \in OBS(k)$ the corresponding S-element is nothing but $R_{kj}X_i - X_kR_{ji}$.

(i) $\Leftrightarrow$ (ii) Since $X_j X_i = \mathbf{LM}(R_{ji})$ for $1 \leq i < j \leq n$, this equivalence follows from CH.II Theorem 3.1 and Proposition 3.4.

(ii) $\Rightarrow$ (iii) This follows from Theorem 1.2.

(iii) $\Rightarrow$ (ii) If each $R_{kj} X_i - X_k R_{ji}$ has a weak Gröbner representation by $\mathcal{G}$, then it follows from Theorem 1.2 that (ii) holds. / $\square$

To realize Theorem 1.5, one may, of course, directly check (iii) by definition, or verify (ii) by means of the very noncommutative version of Buchberger's algorithm given by Mora in [Mor2]. However, to avoid large and tedious noncommutative division procedure, we will see that Berger's q-Jacobi condition [Ber] is quite helpful in the case where the *graded* lexicographic ordering $\geq_{grlex}$ is used (indeed $\geq_{grlex}$ is the most commonly used monomial ordering on free algebras in practice, see CH.II Example (iv) for the definition), though the algebras we are dealing with are not necessarily the q-algebras studied in [Ber] (see Example (vi) in next section). More precisely, Let $k\langle X \rangle$, $A = k\langle X \rangle/I$ be as before, where $I = \langle R_{ji} \mid 1 \leq i < j \leq n \rangle$ with $R_{ji} = X_j X_i - \lambda_{ji} X_i X_j - \{X_j, X_i\}$. Then, for $1 \leq i < j < k \leq n$, the *Jacobi sum* $\mathbf{J}(X_k, X_j, X_i)$ in the sense of [Ber] is defined as follows.

$$
\begin{aligned}
\mathbf{J}(X_k, X_j, X_i) = \ & \{X_k, X_j\} X_i - \lambda_{ki}\lambda_{ji} X_i \{X_k, X_j\} - \\
& - \lambda_{ji} \{X_k, X_i\} X_j + \lambda_{kj} X_j \{X_k, X_i\} + \\
& + \lambda_{kj}\lambda_{ki} \{X_j, X_i\} X_k - X_k \{X_j, X_i\}.
\end{aligned}
$$

1.6. Proposition Let $A = k\langle X \rangle/I$ be as above, and let $\geq_{grlex}$ be the graded lexicographic monomial ordering in $(k\langle X \rangle, \mathcal{B}, \geq_{grlex})$ such that

$$X_n >_{grlex} X_{n-1} >_{grlex} \cdots >_{grlex} X_1, \text{ and}$$
$$\mathbf{LM}(R_{ji}) = X_j X_i \text{ with respect to } \geq_{grlex}, \ 1 \leq i < j \leq n.$$

The following statements are equivalent.

(i) $\mathcal{G} = \{R_{ji} \mid 1 \leq i < j \leq n\}$ forms a Gröbner basis for the ideal $I = \langle \mathcal{G} \rangle \subset k\langle X \rangle$ with respect to $\geq_{grlex}$.

(ii) For $1 \leq i < j < k \leq n$,

$$\mathbf{J}(X_k, X_j, X_i) \in k\text{-span}\left\{ R_{pq} \ \middle| \ 1 \leq q < p \leq n \right\} +$$

$$+ \ k\text{-span}\left\{
\begin{array}{ll}
X_h R_{ji}, R_{ji} X_h, R_{ji} X_k, & 1 \leq h < k, \\
X_h R_{ki}, R_{ki} X_h, R_{ki} X_k, & 1 \leq h < k, \\
X_h R_{kj}, R_{kj} X_m, & 1 \leq h < k, 1 \leq m < i
\end{array}
\right\}$$

Proof From the defining relations we derive

$$
\begin{aligned}
\{X_k, \ X_j\} &= X_k X_j - \lambda_{kj} X_j X_k - R_{kj}, \\
\{X_k, \ X_i\} &= X_k X_i - \lambda_{ki} X_i X_k - R_{ki}, \\
\{X_j, \ X_i\} &= X_j X_i - \lambda_{ji} X_i X_j - R_{ji}.
\end{aligned}
$$

It follows that

$$\mathbf{J}(X_k, X_j, X_i) = X_k R_{ji} - R_{kj} X_i + \\
+ \lambda_{ji} R_{ki} X_j - \lambda_{kj} X_j R_{ki} - \lambda_{kj} \lambda_{ki} R_{ji} X_k + \lambda_{ki} \lambda_{ji} X_i R_{kj},$$

which is obviously contained in I, and consequently

$$R_{kj} X_i - X_k R_{ji} = \lambda_{ji} R_{ki} X_j - \lambda_{kj} X_j R_{ki} - \lambda_{kj} \lambda_{ki} R_{ji} X_k + \lambda_{ki} \lambda_{ji} X_i R_{kj} - \\
- \mathbf{J}(X_k, X_j, X_i).$$

Noticing the ordering $X_n >_{grlex} X_{n-1} >_{grlex} \cdots >_{grlex} X_1$, the equivalence (i) $\Leftrightarrow$ (ii) is clear now by Theorem 1.5.   □

Applications of Theorem 1.5 and Proposition 1.6 are given in the next section.

2. Quadric Solvable Polynomial Algebras

Based on the results obtained in §1, in this section we formulate and construct quadric solvable polynomial algebras, which form the practical working basis for later chapters.

Let $A = k[a_1, ..., a_n]$ be a *solvable polynomial algebra* in the sense of CH.II Definition 7.3, with the associated (left) admissible system $(A, \mathcal{B}, \succeq_{gr})$, where

$$\mathcal{B} = \{a_1^{\alpha_1} a_2^{\alpha_2} \cdots a_n^{\alpha_n} \mid (\alpha_1, ..., \alpha_n) \in I\!N^n\}$$

is the k-basis consisiting of standard monomials and $\succeq_{gr}$ is a *graded monomial ordering*. It follows that the generators of A satisfy *only* quadric relations, that is, for $1 \le i < j \le n$,

$$(*) \qquad a_j a_i = \lambda_{ji} a_i a_j + \sum_{k \le \ell} \lambda_{ji}^{k\ell} a_k a_\ell + \sum \lambda_h a_h + c_{ji},$$

where λ_{ji}, $\lambda_{ji}^{k\ell}$, λ_h, $c_{ji} \in k$, and $\lambda_{ji} \ne 0$. This leads to the following specific class of solvable polynomial algebras.

2.1. Definition We call the above solvable polynomial algebra A a *quadric solvable polynomial algebra*.
If $\sum \lambda_h a_h + c_{ji} = 0$ in the formula $(*)$, we call A a *quadratic* solvable polynomial algebra; if $\sum \lambda_{ji}^{k\ell} a_k a_\ell = 0$ in the formula $(*)$, we call A a *linear* solvable polynomial algebra; if $\sum \lambda_{ji}^{k\ell} a_k a_\ell + \sum \lambda_h a_h + c_{ji} = 0$ in the formula $(*)$, we call A a *homogeneous* solvable polynomial algebra.
In particular, if k, $\ell < j$ in the formula $(*)$ whenever $\lambda_{ji}^{k\ell} \ne 0$, then we call A a *tame* quadric solvable polynomial algebra.

From the definition it is clear that any linear or homogeneous solvable polynomial algebra is trivially a tame quadric solvable polynomial algebra. More generally, if $A = k[x_1; \sigma_1, \delta_1][x_2; \sigma_2, \delta_2] \cdots [x_n; \sigma_n, \delta_n]$ is an iterated skew polynomial algebra starting with the ground field k (see CH.II §1 Example (ii)), where each σ_i is an algebra automorphism and the δ_i is a σ_i-derivation, such that $\sigma_j(x_i) = \lambda_{ji} x_i$ with $\lambda_{ji} \in k - \{0\}$ for all $j > i$, and that $\delta_j(x_i) = p_{ji}$ is a linear combination of the monomials $x_1^{\alpha_1} \cdots x_\ell^{\alpha_\ell}$ with $\ell < j$ and $\alpha_1 + \cdots + \alpha_\ell \le 2$, for all $j > i$, then A is a tame quadric solvable polynomial algebra with respect to $x_n >_{grlex} x_{n-1} >_{grlex} \cdots >_{grlex} x_1$. Below we construct quadric solvable polynomial algebras in which some do not seem having a specific iterated skew polynomial algebra structure starting with the ground field k (at least by first glance at the given defining relations). And we also show that tame quadric solvable polynomial algebras are completely constructable.

Let $A = k[a_1, ..., a_n]$ be a quadric solvable polynomial k-algebra with $\succeq_{gr}$. Then, since the generators satisfy the relations as given in previous (*), and since $\mathcal{B} = \{a_1^{\alpha_1} \cdots a_n^{\alpha_n} \mid (\alpha_1, ..., \alpha_n) \in \mathbb{N}^n\}$ forms a k-basis for A, we have $A \cong k\langle X \rangle / I$, where $k\langle X \rangle = k\langle X_1, ..., X_n \rangle$ is the free associative k-algebra over $X = \{X_1, ..., X_n\}$ and I is the ideal of $k\langle X \rangle$ generated by

$$R_{ji} = X_j X_i - \lambda_{ji} X_i X_j - \sum_{k \le \ell} \lambda_{ji}^{k\ell} X_k X_\ell - \sum \lambda_h X_h - c_{ji}, \ 1 \le i < j \le n,$$

or in other words, $\{R_{ji} \mid 1 \le i < j \le n\}$ is a set of defining relations for A. This observation opens to us the door for recognizing and constructing quadric solvable polynomial algebras in an algorithmic way, that is, we have the following consequence of Theorem 1.5.

2.2. Proposition Consider the k-algebra $A = k\langle X \rangle / I$, where I is the ideal of $k\langle X \rangle$ generated by the quadric defining relations

$$R_{ji} = X_j X_i - \lambda_{ji} X_i X_j - \sum \lambda_{ji}^{k\ell} X_k X_\ell - \sum \lambda_h X_h - c_{ji}, \ 1 \le i < j \le n,$$

where $\lambda_{ji}, \lambda_{ji}^{k\ell}, \lambda_h, c_{ji} \in k$. Suppose that
(1) $\lambda_{ji} \ne 0, 1 \le i < j \le n$, and
(2) one of the following conditions is satisfied whenever $\lambda_{ji}^{k\ell} \ne 0$:
    (a) $k = \ell$ and $k, \ell < j$.
    (b) $k \ne \ell$ and $k, \ell \le j$, where $k = j$ implies $\ell < i$ and $\ell = j$ implies $k < i$.
Then $A = k[a_1, ..., a_n]$ is a quadric solvable polynomial algebra with respect to $a_n >_{grlex} a_{n-1} >_{grlex} \cdots >_{grlex} a_1$, where each a_i is the image of X_i in A, if and only if $\{R_{ji} \mid 1 \le i < j \le n\}$ forms a Gröbner basis in $k\langle X \rangle$ with respect to $X_n >_{grlex} X_{n-1} >_{grlex} \cdots >_{grlex} X_1$.

Proof Suppose that $\{R_{ji} \mid 1 \le i < j \le n\}$ forms a Gröbner basis in $k\langle X \rangle$ with respect to $X_n >_{grlex} X_{n-1} >_{grlex} \cdots >_{grlex} X_1$. Since by the assumption (2) we have $\mathbf{LM}(R_{ji}) = X_j X_i, \ 1 \le i < j \le n$, it follows from Theorem 1.5 that

$\mathcal{B} = \{a_1^{\alpha_1} a_2^{\alpha_2} \cdots a_n^{\alpha_n} \mid (\alpha_1, ..., \alpha_n) \in I\!\!N^n\}$ forms a k-basis for A. Now one checks directly that the assumption (1)–(2) and the defining relations together make A into a quadric solvable polynomial algebra with respect to $a_n >_{grlex} a_{n-1} >_{grlex} \cdots >_{grlex} a_1$. The converse is clear by Theorem 1.5. □

An immediate application of Proposition 2.2 is to show that tame quadric solvable polynomial algebras are completely constructable.

2.3. Proposition Let $A = k[a_1, ..., a_n]$ be a tame quadric solvable polynomial algebra in the sense of Definition 2.1, that is, the generators of A satisfy

$$a_j a_i = \lambda_{ji} a_i a_j + \sum_{k \le \ell < j} \lambda_{ji}^{k\ell} a_k a_\ell + \sum \lambda_h a_h + c_{ji}, \ 1 \le i < j \le n.$$

Setting on the free algebra $k\langle X \rangle = k\langle X_1, ..., X_n \rangle$ the graded monomial ordering $X_n >_{grlex} X_{n-1} >_{grlex} \cdots >_{grlex} X_1$, A is isomorphic to the tame quadric solvable polynomial algebra $k[b_1, ..., b_n] = k\langle X \rangle / I$ with respect to $b_n >_{grlex} b_{n-1} >_{grlex} \cdots >_{grlex} b_1$, where each b_i is the residue of X_i (modulo I) and I is generated by the relations

$$R_{ji} = X_j X_i - \lambda_{ji} X_i X_j - \sum_{k \le \ell < j} \lambda_{ji}^{k\ell} X_k X_\ell - \sum \lambda_h X_h - c_{ji}, \ 1 \le i < j \le n.$$

Proof First note that, with the given data of A, $A \cong k[b_1, ..., b_n] = k\langle X \rangle / I$ as k-algebras. That $k[b_1, ..., b_n]$ is a tame quadric solvable polynomial algebra with respect to $b_n >_{grlex} b_{n-1} >_{grlex} \cdots >_{grlex} b_1$ follows from Theorem 1.5 and Proposition 2.2. □

We are ready to construct quadric solvable polynomial algebras by using Proposition 2.2 and Proposition 1.6. In the examples given below, notation are maintained as before. Moreover, by abusing language, some examples will be called "deformations" of certain well-known algebras.

Example (i) Let $X_2 X_1 - q X_1 X_2 - a X_1^2 - b X_1 - c X_2 - d = R_{21} \in k\langle X_1, X_2 \rangle$, where $q, a, b, c, d \in k$. By Corollary 1.3, $\{R_{21}\}$ is a Gröbner basis in $k\langle X_1, X_2 \rangle$ with respect to $X_2 >_{grlex} X_1$. Thus, If $q \ne 0$, then the algebra $A = k\langle X_1, X_2 \rangle / \langle R_{21} \rangle$ is a tame quadric solvable polynomial algebra with respect to $\ge_{grlex}$ (indeed this is a skew polynomial algebra). If $q \ne 0$, $a = b = c = 0$, and $d = 1$, then $A = A_1(q)$, the additive analogue of the first weyl algebra. One sees that here we have got all 2-dimensional quadric solvable polynomial algebras with respect to $X_2 >_{grlex} X_1$.

(ii) Deformations of $U(sl_2)$.
Let $U(sl_2)$ be the enveloping algebra of the 3-dimensional Lie algebra $sl_2 = kx \oplus ky \oplus kz$ defined by the bracket product: $[x, y] = z$, $[z, x] = 2x$, $[z, y] = -2y$.

This example provides quadric solvable polynomial algebras which are deformations of $U(sl_2)$.

Let $k\langle X_1, X_2, X_3\rangle$ be the free k-algebra over $\{X_1, X_2, X_3\}$, and $k\langle X_1, X_2, X_3\rangle / I = A$ where I is the two-sided ideal generated by the defining relations

$$R_{21} = X_2X_1 - \alpha X_1X_2 - \gamma X_2 - F_{21},$$
$$R_{31} = X_3X_1 - \frac{1}{\alpha}X_1X_3 + \frac{\gamma}{\alpha}X_3 - F_{31},$$
$$R_{32} = X_3X_2 - \beta X_2X_3 - F(X_1) - F_{32},$$

where

$$\alpha \neq 0, \ \beta, \ \gamma \in k, \quad F(X_1) \in k\text{-span}\{X_1^2, X_1, 1\},$$
$$F_{21}, \ F_{31}, \ F_{32} \in k\langle X_1, X_2, X_3\rangle.$$

If $\alpha = \beta = 1$, $\gamma = 2$, $F(X_1) = X_1$, and $F_{21} = F_{31} = F_{32} = 0$, then $A = U(sl_2)$. Moreover, in the case where $F_{21} = F_{31} = F_{32} = 0$, the family of algebras constructed above includes many well-known deformations of $U(sl_2)$, e.g., Woronowicz's deformation of $U(sl_2)$ [Wor], Witten's deformation of $U(sl_2)$ [Wit], Le Bruyn's conformal sl_2 enveloping algebra [Le2], Smith's deformation of $U(sl_2)$ where the dominant polynomial $f(t)$ has degree ≤ 2 [Sm2], Benkart-Roby's down-up algebra in which $\beta \neq 0$ (cf. [KMP], [CM]).

Set on $k\langle X_1, X_2, X_3\rangle$ the monomial ordering $X_3 >_{grlex} X_2 >_{grlex} X_1$. Then the only Jacobi sum determined by the defining relations of A with respect to the fixed ordering on generators is

$$
\begin{aligned}
\mathbf{J}(X_3, X_2, X_1) &= \{X_3, X_2\}X_1 - \lambda_{31}\lambda_{21}X_1\{X_3, X_2\}- \\
&\quad -\lambda_{21}\{X_3, X_1\}X_2 + \lambda_{32}X_2\{X_3, X_1\}+ \\
&\quad +\lambda_{32}\lambda_{31}\{X_2, X_1\}X_3 - X_3\{X_2, X_1\} \\
&= (f(X_1) + F_{32})X_1 - \frac{1}{\alpha} \cdot \alpha X_1(f(X_1) + F_{32})- \\
&\quad -\alpha\left(-\frac{\gamma}{\alpha}X_3 + F_{31}\right)X_2 + \beta X_2\left(-\frac{\gamma}{\alpha}X_3 + F_{31}\right) + \\
&\quad +\beta \cdot \frac{1}{\alpha}(\gamma X_2 + F_{21})X_3 - X_3(\gamma X_2 + F_{21}) \\
&= F_{32}X_1 - X_1F_{32} - \alpha F_{31}X_2 + \beta X_2F_{31} + \frac{\beta}{\alpha}F_{21}X_3 - X_3F_{21}.
\end{aligned}
$$

Write

$$\mathcal{F} = F_{32}X_1 - X_1F_{32} - \alpha F_{31}X_2 + \beta X_2F_{31} + \frac{\beta}{\alpha}F_{21}X_3 - X_3F_{21}.$$

By Proposition 1.6, if

$$\mathbf{LM}(R_{ji}) = X_jX_i \text{ w.r.t. } \geq_{grlex}, \ 1 \leq i < j \leq 3, \text{ and}$$

$$\mathcal{F} \in k\text{-span}\left\{
\begin{array}{l}
R_{21}, R_{31}, R_{32}, \\
X_1R_{21}, R_{21}X_1, X_2R_{21}, R_{21}X_2, R_{21}X_3, \\
X_1R_{31}, R_{31}X_1, X_2R_{31}, R_{31}X_2, R_{31}X_3, \\
X_1R_{32}, X_2R_{32}
\end{array}
\right\},$$

then $\{R_{21}, R_{31}, R_{32}\}$ forms a Gröbner basis with respect to $X_3 >_{grlex} X_2 >_{grlex} X_1$. Below we consider two cases:

Case I. Input in the defining relations of A the data

(D1)
$$\begin{cases} \alpha = \beta \neq 0, \ \gamma, \ \mu, \ q, \ \varepsilon, \ \xi, \ \lambda, \ \eta_{32} \in k, \\ F(X_1) \in k\text{-span}\{X_1^2, \ X_1, \ 1\}, \\ G(X_2) \in k\text{-span}\{X_2^2, \ X_2, \ 1\}, \\ H(X_3) \in k\text{-span}\{X_3, \ 1\}, \\ F_{21} = \mu X_1^2 + q X_1 + H(X_3), \\ F_{31} = \varepsilon(X_1 X_2 + X_2 X_1) - \xi X_1^2 + \lambda X_1 + G(X_2), \\ F_{32} = (\mu(X_1 X_3 + X_3 X_1) - \varepsilon \alpha X_2^2) + \xi \alpha(X_1 X_2 + X_2 X_1) \\ \qquad - \lambda \alpha X_2 + q X_3 + \eta_{32}. \end{cases}$$

Clearly, in this case we have $\mathbf{LM}(R_{ji}) = X_j X_i$, $1 \leq i < j \leq 3$, and the conditions of Proposition 2.2 are satisfied. Moreover, a direct verification shows that

$$\begin{aligned} F_{32}X_1 - X_1 F_{32} &= \mu(X_3 X_1^2 - X_1^2 X_3) + \varepsilon\alpha(X_1 X_2^2 - X_2^2 X_1) + \\ &\quad + \xi\alpha(X_2 X_1^2 - X_1^2 X_2) + \lambda\alpha(X_1 X_2 - X_2 X_1) + \\ &\quad + q(X_3 X_1 - X_1 X_3), \\ -\alpha F_{31}X_2 + \alpha X_2 F_{31} &= \varepsilon\alpha(X_2^2 X_1 - X_1 X_2^2) + \xi\alpha(X_1^2 X_2 - X_2 X_1^2) + \\ &\quad + \lambda\alpha(X_2 X_1 - X_1 X_2), \\ F_{21}X_3 - X_3 F_{21} &= \mu(X_1^2 X_3 - X_3 X_1^2) + q(X_1 X_3 - X_3 X_1), \end{aligned}$$

and concequently, $J(X_3, X_2, X_1) = \mathcal{J} = 0$. By Proposition 2.2, A is a quadric solvable polynomial algebra.

Case II. Input in the defining relations of A the data

(D2)
$$\begin{cases} \alpha = \beta \neq 0, \ \gamma, \ q, \ \varepsilon, \ \xi, \ \lambda, \ \eta_{32} \in k, \\ F(X_1) \in k\text{-span}\{X_1^2, \ X_1, \ 1\}, \\ G(X_2) \in k\text{-span}\{X_2^2, \ X_2, \ 1\}, \\ H(X_3) \in k\text{-span}\{X_3, \ 1\}, \\ F_{21} = q X_1 + H(X_3), \\ F_{31} = \varepsilon(X_1 X_2 + X_2 X_1) - \xi X_1^2 + \lambda X_1 + G(X_2), \\ F_{32} = -\varepsilon\alpha X_2^2 + \xi\alpha(X_1 X_2 + X_2 X_1) - \lambda\alpha X_2 + q X_3 + \eta_{32}, \end{cases}$$

As with the data (D1), one checks that in this case we also have $J(X_3, X_2, X_1) = 0$. So A is a tame quadric solvable polynomial algebra.

By taking the residues (modulo I) in the above cases, one may obtain a set $\{F_{21}, F_{31}, F_{32}\}$ in which each member F_{ij} is a linear combination of standard monomials. •

(iii) Non-polynomial central extensions of the deformations of $U(sl_2)$.
These are the 4-dimensional algebras defined by the relations from the free

algebra $k\langle X_1, X_2, X_3, X_4 \rangle$

$$R_{21} = X_2 X_1 - \alpha X_1 X_2 - \gamma X_2 - F_{21} - K_{21}$$
$$R_{31} = X_3 X_1 - \tfrac{1}{\alpha} X_1 X_3 + \tfrac{\gamma}{\alpha} X_3 - F_{31} - K_{31}$$
$$R_{32} = X_3 X_2 - \alpha X_2 X_3 - F(X_1) - F_{32} - K_{32}$$
$$R_{41} = X_4 X_1 - X_1 X_4,$$
$$R_{42} = X_4 X_2 - X_2 X_4,$$
$$R_{43} = X_4 X_3 - X_3 X_4,$$

where $K_{21}, K_{31}, K_{32} \in k\text{-span}\{X_4,\ 1\}$ and $\{\alpha, \gamma, F(X_1), F_{21}, F_{31}, F_{32}, F_{32}\}$ is taken either from (**D1**) or from (**D2**) in Example (ii). Since the only possible nonzero Jacobi sums determined by the above relations with respect to $X_4 >_{grlex} X_3 >_{grlex} X_2 >_{grlex} X_1$ are given by

$$\mathbf{J}(X_3, X_2, X_1) = K_{32} X_1 - X_1 K_{32} - \alpha K_{31} X_2 + \alpha X_2 K_{31} + K_{21} X_3 - X_3 K_{21},$$
$$\mathbf{J}(X_4, X_3, X_2) = F(X_1) X_4 - X_4 F(X_1) + F_{32} X_4 - X_4 F_{32},$$
$$\mathbf{J}(X_4, X_3, X_1) = \tfrac{\gamma}{\alpha}(X_4 X_3 - X_3 X_4) + F_{31} X_4 - X_4 F_{31},$$
$$\mathbf{J}(X_4, X_2, X_1) = \gamma(X_2 X_4 - X_4 X_2) + F_{21} X_4 - X_4 F_{21},$$

it can be further checked that these sums have weak Gröbner representations by $\{R_{41}, R_{42}, R_{43}\}$. Thus, Proposition 2.2 and Proposition 1.6 hold. Hence, the algebras defined by the relations given above are quadric solvable polynomial algebras.

(iv) Deformations of $A_n(k)$.

Let $A_n(k)$ be the nth Weyl algebra over k. This example provides quadric solvable polynomial algebras which are deformations of $A_n(k)$.

Set on the free algebra $k\langle Y, X \rangle = k\langle Y_n, ..., Y_1, X_n, ..., X_1 \rangle$ the monomial ordering

$$Y_n >_{grlex} X_n >_{grlex} Y_{n-1} >_{grlex} X_{n-1} >_{grlex} \cdots >_{grlex} Y_1 >_{grlex} X_1.$$

Consider the algebra A with defining relations

$$
\begin{array}{ll}
H_{ji} = X_j X_i - X_i X_j, & 1 \le i < j \le n, \\
\widetilde{H}_{ji} = X_j Y_i - Y_i X_j, & 1 \le i < j \le n, \\
G_{ji} = Y_j Y_i - Y_i Y_j, & 1 \le i < j \le n, \\
\widetilde{G}_{ji} = Y_j X_i - X_i Y_j, & 1 \le i < j \le n \\
R_{jj} = Y_j X_j - q_j X_j Y_j - F_{jj}, & 1 \le j \le n,
\end{array}
$$

where $q_j \in k$, $F_{jj} \in k\langle Y, X \rangle$. If in the defining relations $F_{jj} = 1$ for $1 \le i < j \le n$, then the additive analogue $A_n(q_1, ..., q_n)$ of the Weyl algebra (CH.I §5 Example (ii)) is recaptured.

A direct verification shows that the only possible nonzero Jacobi sums determined by the defining relations and the ordering given on generators are

$$
\begin{array}{ll}
\mathbf{J}(Y_j, X_j, X_i) = F_{jj} X_i - X_i F_{jj}, & 1 \le i < j \le n, \\
\mathbf{J}(Y_j, X_j, Y_i) = F_{jj} Y_i - Y_i F_{jj}, & 1 \le i < j \le n, \\
\mathbf{J}(Y_k, Y_j, X_j) = F_{jj} Y_k - Y_k F_{jj}, & 1 \le j < k \le n, \\
\mathbf{J}(X_k, Y_j, X_j) = F_{jj} X_k - X_k F_{jj}, & 1 \le j < k \le n.
\end{array}
$$

For $1 \le j \le n$, at least if

$$F_{jj} \in k\text{-span}\{X_j^2, \ X_j, \ Y_j, \ 1\},$$

then all conditions of Proposition 2.2 and Proposition 1.6 are satisfied, and one checks that all Jacobi sums have weak Gröbner representations. It follows that A is a tame quadric solvable polynomial algebra with $\ge_{grlex}$ in the case where all $q_j \ne 0$.

(v) Deformations of the Heisenberg enveloping algebra.
Let $k\langle X \cup Z \cup Y \rangle$ be the free k algebra over $\{X_n, ..., X_1, Z_n, ..., Z_1, Y_n, ..., Y_1\} = X \cup Z \cup Y$, $A = k\langle X \cup Z \cup Y \rangle / I$, where I is the ideal generated by the defining relations

$$
\begin{aligned}
R_{ji}^{x} &= X_j X_i - X_i X_j, & 1 &\le i < j \le n,\\
R_{ji}^{y} &= Y_j Y_i - Y_i Y_j, & 1 &\le i < j \le n,\\
R_{ji}^{z} &= Z_j Z_i - Z_i Z_j, & 1 &\le i < j \le n,\\
R_{ji}^{zy} &= Z_j Y_i - \lambda_i^{\delta_{ji}} Y_i Z_j, & 1 &\le i, \ j \le n,\\
R_{ji}^{xz} &= X_j Z_i - \mu_i^{\delta_{ji}} Z_i X_j, & 1 &\le i, \ j \le n,\\
R_{ji}^{xy} &= X_j Y_i - Y_i X_j, & i &\ne j,\\
R_{jj}^{xy} &= X_j Y_j - q_j Y_j X_j - F_{jj}, & 1 &\le i \le n,
\end{aligned}
$$

where $\lambda_i, \mu_i, q_j \in k$, $F_{jj} \in k\langle X \cup Z \cup Y \rangle$. If we take $q_j = 1$ $Z_j = Z$ and $F_{jj} = Z$, $1 \le j \le n$, then the enveloping algebra of $2n + 1$-dimensional Heisenberg Lie algebra is recovered. In the case where $\lambda_i = \mu_i = q \ne 0$, $q_j = q^{-1}$, and $F_{jj} = z_j$, we recover the q-Heisenberg algebra (CH.I §5 Example (v)).
Set the monomial ordering

$$X_n >_{grlex} \cdots >_{grlex} X_1 >_{grlex} Z_n >_{grlex} \cdots >_{grlex} Z_1 >_{grlex}$$
$$>_{grlex} Y_n >_{grlex} \cdots >_{grlex} Y_1.$$

Then a direct verification shows that the only possible nonzero Jacobi sums determined by the defining relations and the ordering given on generators are

$$
\begin{aligned}
\mathbf{J}(X_k, X_j, Y_j) &= F_{jj} X_k - X_k F_{jj}, & 1 &\le j < k \le n,\\
\mathbf{J}(X_k, X_j, Y_k) &= -F_{kk} X_j + X_j F_{kk}, & 1 &\le j < k \le n,\\
\mathbf{J}(X_k, Z_j, Y_k) &= -F_{kk} Z_j + Z_j F_{kk}, & 1 &\le k, \ j \le n,\\
\mathbf{J}(X_k, Y_k, Y_j) &= F_{kk} Y_j - Y_j F_{kk}, & 1 &\le j < k \le n,\\
\mathbf{J}(X_j, Y_k, Y_j) &= -F_{jj} Y_k + Y_k F_{jj}, & 1 &\le j < k \le n.
\end{aligned}
$$

It can be further checked that, for $1 \le j \le n$, at least if

$$F_{jj} \in k\text{-span}\{Z_j^2, \ Z_j, \ Y_j^2, \ Y_j, \ X_j, \ 1\},$$

then all conditions of Proposition 2.2 and Proposition 1.6 are satisfied, and all Jacobi sums have weak Gröbner representations by the defining relations. It follows that A is a tame quadric solvable polynomial algebra with $\ge_{grlex}$ in the case where all $q_j \ne 0$.

(vi) Berger's q-enveloping algebras.

Recall from [Ber] that a q-algebra $A = k[a_1, ..., a_n]$ over a *commutative ring k* is defined by the quadric relations

$$R_{ji} = X_j X_i - q_{ji} X_i X_j - \{X_j, X_i\}, \ 1 \leq i < j \leq n, \text{ where } q_{ji} \in k,$$
$$\text{and } \{X_j, X_i\} = \sum \alpha_{ji}^{k\ell} X_k X_\ell + \sum \alpha_h X_h + c_{ji}, \ \alpha_{ji}^{k\ell}, \alpha_h, c_{ji} \in k,$$
$$\text{satisfying if } \alpha_{ji}^{kl} \neq 0, \text{ then } i < k \leq \ell < j, \text{ and } k - i = j - \ell.$$

Define two k-subspaces of the free algebra $k\langle X_1, ..., X_n \rangle$

$$\mathcal{E}_1 = k\text{-Span}\left\{ R_{ji} \ \middle| \ n \geq j > i \geq 1 \right\},$$

$$\mathcal{E}_2 = k\text{-Span}\left\{ X_i R_{ji}, \ R_{ji} X_i, \ X_j R_{ji}, \ R_{ji} X_j \ \middle| \ n \geq j > i \geq 1 \right\}.$$

For $1 \leq i < j < k \leq n$, if every Jacobi sum $\mathbf{J}(X_k, X_j, X_i)$ is contained in $\mathcal{E}_1 + \mathcal{E}_2$, then A is called a q-*enveloping* algebra with respect to the natural total ordering $a_n > a_{n-1} > \cdots > a_1$. A q-enveloping algebra is said to be *invertible* if in the defining relations all coefficients q_{ji} are invertible, $1 \leq i < j \leq n$. In [Ber] the q-PBW theorem was obtained for q-enveloping algebras, that is, the set of standard monomials $\{a_1^{\alpha_1} \cdots a_n^{\alpha_n} \mid (\alpha_1, ..., \alpha_n) \in I\!N^n\}$ forms a k-basis for a q-enveloping algebra A. Clearly, if we set the monomial ordering $X_n >_{grlex} X_{n-1} >_{grlex} \cdots >_{grlex} X_1$, then the defining relations of a q-algebra A satisfy

$$\mathbf{LM}(R_{ji}) = X_j X_i, \ 1 \leq i < j \leq n \text{ and}$$
$$k, \ell < j \text{ in } \sum \alpha_{ji}^{k\ell} X_k X_\ell \text{ whenever } \alpha_{ji}^{k\ell} \neq 0.$$

Hence, by Proposition 1.6 and Proposition 2.2, the defining relations of a q-enveloping algebra form a Gröbner basis in $k\langle X_1, ..., X_n \rangle$; if furthermore A is an invertible q-enveloping algebra then A is a tame quadric solvable polynomial algebra (indeed, A is an iterated skew polynomial algebra by the proof of Theorem 2.9.4 in [Ber]).

We observe that the conditions $i < k \leq \ell < j$ and $k - i = j - \ell$ in the definition of a q-algebra are not necessarily satisfied by a quadric solvable polynomial algebra, or more generally, a quadric algebra characterizied by Proposition 1.6 is not necessarily a q-enveloping algebra in the sense of [Ber].

Remark In the end of first part of [LWZ], it was pointed out that a q-enveloping algebra over a field k is generally not a solvable polynomial algebra with respect to $\geq_{grlex}$. This is, of course, not true for invertible q-enveloping algebras, as argued in the above example (vi). The author takes this place to correct that incorrect remark.

3. Associated Homogeneous Defining Relations of Algebras

Let k be any field, let $k\langle X\rangle$ be the free k-algebra over $X = \{X_i\}_{i\in\Lambda}$, and let $S = \langle X\rangle$ be the free semigroup generated by $X = \{X_i\}_{i\in\Lambda}$. From CH.I §4 Example (i) and (iv) we retain that $k\langle X\rangle$ has the natural gradation defined by the degree of words in S, i.e., $k\langle X\rangle = \oplus_{p\in\mathbb{N}}k\langle X\rangle_p$ with $k\langle X\rangle_p = \{\sum_{d(w)=p}c_w w \mid c_w \in k,\ w\in S\}$, and that the standard filtration (or the grading filtration) $Fk\langle X\rangle$ on $k\langle X\rangle$ is given by the k-subspaces:

$$F_p k\langle X\rangle = \oplus_{i\leq p}k\langle X\rangle_i, \quad p\in\mathbb{N}.$$

If I is a two-sided ideal of $k\langle X\rangle$ and $A = k\langle X\rangle/I$, then $Fk\langle X\rangle$ induces a filtration FA on A:

$$F_p A = (F_p k\langle X\rangle + I)/I, \quad p\in\mathbb{N}.$$

Indeed, FA coincides with the standard filtration on the k-algebra $A = k\langle X\rangle/I = k[a_i]_{i\in\Lambda}$ where each a_i is the image of X_i in $k\langle X\rangle/I$.

If we consider the associated graded algebra $G(A) = \oplus_{p\in\mathbb{N}}(F_p A/F_{p-1}A)$ of A and the Rees algebra $\widetilde{A} = \oplus_{p\in\mathbb{N}}F_p A$ of A, then it follows from CH.I Proposition 4.4 that $G(A)$ and $\widetilde{A}$ are determined by I^*, where I^* is the homogenization ideal of I in the polynomial ring $k\langle X\rangle[t]$ in commuting variable t, namely,

$$(*) \qquad \begin{cases} \widetilde{A} \cong k\langle X\rangle[t]/I^*, \text{ and} \\[2mm] G(A) \cong k\langle X\rangle[t]/(\langle t\rangle + I^*), \end{cases}$$

where $\langle t\rangle$ denotes the ideal of $k\langle X\rangle[t]$ generated by t.

With notation as above, let $\{f_j\}_{j\in J}$ be a set of defining relations of the k-algebra $A = k\langle X\rangle/I$, i.e., the two-sided ideal I is generated by $\{f_j\}_{j\in J}$. Let FA be the standard filtration on A, and let $G(A)$ and $\widetilde{A}$ be the associated graded algebra and Rees algebra of A with respect to FA, respectively. In view of the above $(*)$, it is natural to ask the following question:

Question Can we determine the defining relations of $G(A)$ and $\widetilde{A}$ in terms of the defining relations of A?

Inspired by a remarkable application of Gröbner basis in algebraic geometry, in this section we give an (even stronger) positive answer to the above question. To be convenient, we recall from (e.g., [CLO']) the following result.

Let k be an algebraically closed field of characteristic 0, and $k[x_1, ..., x_n]$ the *commutative* polynomial k-algebra in n variables. Let I be an ideal of $k[x_1, ..., x_n]$, and let I^* denote the homogenization ideal of I in $k[x_0, x_1, ..., x_n]$ with respect to x_0. Then the projective algebraic set $V(I^*)$ defined by I^* in the projective n-space $\mathbf{P}_k^n$ gives the projective closure of the affine algebraic set $V(I)$ defined by I in the affine n-space $\mathbf{A}_k^n$. The following result tells us that, using the Gröbner basis method, the defining equations of the projective closure $V(I^*)$ of the affine algebraic set $V(I)$ can be determined from that of $V(I)$, or equivalently, the defining relations of the graded k-algebra $k[x_0, x_1, ..., x_n]/I^*$ can be determined from that of $k[x_1, ..., x_n]/I$.

- If $\mathcal{G} = \{g_1, ..., g_s\}$ is a Gröbner basis for I with respect to a *graded monomial ordering* in $k[x_1, ..., x_n]$, then $\mathcal{G}^* = \{g_1^*, ..., g_s^*\}$ is a Gröbner basis for $I^* \subset k[x_0, x_1, ..., x_n]$, where g_i^* is the homogenization of g_i in $k[x_0, x_1, ..., x_n]$.

Before studying the question posed above, let us also recall some well known examples (see CH.I §5).

Example (i) Let $\mathbf{g} = kx_1 \oplus \cdots \oplus kx_n$ be an n-dimensional Lie algebra over k with $[x_i, x_j] = \sum_{h=1}^n \lambda_{ij}^h x_h$, and let $A = U(\mathbf{g})$ be the enveloping algebra of $\mathbf{g}$ with the standard filtration $FU(\mathbf{g})$. Then by the PBW theorem we know that $G(U(\mathbf{g}))$ is, as a graded k-algebra, isomorphic to the polynomial k-algebra in n variables.

(ii) Let $A = A_n(k) = k[x_1, ..., x_n, y_1, ..., y_n]$ be the nth Weyl algebra over k as stated in CH.I §5. Then it is well known that, with respect to the standard filtration (or Bernstein filtration) on $A_n(k)$, $G(A_n(k))$ is, as a graded k-algebra, isomorphic to the polynomial k-algebra in $2n$ variables.

Note that in both examples (i) and (ii) the proof of the fact about $G(A)$ is nontrivial in the literature (e.g., [Bj], [Dix]).

(iii) Let $\mathbf{g}$ and $A = U(\mathbf{g})$ be as in (i). Related to the study of quantum groups, S.P. Smith proposed in [Sm1] the quadratic algebra $\widetilde{U}(\mathbf{g})$ which is generated by $x_0, x_1, ..., x_n$ where x_0 is taken to be central and the remaining defining relations are $[x_i, x_j] = \sum_{h=1}^n c_{ij,h} x_h x_0$. In [LeS] and [LeV], this algebra was called the *homogenized enveloping algebra*. Observe that $\widetilde{U}(\mathbf{g})$ looks very like the Rees algebra of $U(\mathbf{g})$, namely, there is $\widetilde{U}(\mathbf{g})/\langle 1 - x_0 \rangle \widetilde{U}(\mathbf{g}) \cong U(\mathbf{g})$. (We will see in the end of this section that $\widetilde{U}(\mathbf{g})$ is exactly the Rees algebra of $U(\mathbf{g})$.) A general interpretation on the homogenized algebra of a given algebra is given after Theorem 3.7 below.

From the above examples one might expect that for a k-algebra A with standard filtration FA, the defining relations of $G(A)$, respectively $\widetilde{A}$, may be given by simply taking the highest degree homogeneous part of the defining relations of A, respectively by simply taking the homogenizations of the defining relations of A in the sense of CH.I §4. As indicated by the following examples, however, (even in the commutative case) the question we posed above is not so trivial to answer in general.

Example (iv) Consider $I = \langle f_1, f_2 \rangle = \langle x_2 - x_1^2, x_3 - x_1^3 \rangle$, the ideal of the affine twisted cubic in $\mathbb{R}^3$. If we homogenize f_1, f_2, then we get the ideal $J = \langle x_2 x_0 - x_1^2, x_3 x_0^2 - x_1^3 \rangle$ in $\mathbb{R}[x_0, x_1, x_2, x_3]$. One may directly check that for $f_3 = f_2 - x_1 f_1 = x_3 - x_1 x_2 \in I$, $f_3^* = x_3 x_0 - x_1 x_2 \notin J$, i.e., $J \neq I^*$.

(v) Let $k\langle X \rangle$ be the free k-algebra generated by $\{X_1, X_2, X_3\}$, and let $f = 2X_3 X_2 X_1 - 3X_1 X_3^2$, $g_{12} = X_2 X_1 - X_1 X_2$, $g_{13} = X_3 X_1 - X_1 X_3$, $g_{23} = X_3 X_2 - X_2 X_3$. Considering the two-sided ideal $I = \langle f, g_{12}, g_{13}, g_{23} \rangle$, then it can be directly verified that $h = -3X_1 X_3^2 + 2X_1 X_2 X_3 = f - 2X_3 g_{12} + 2g_{13} x_2 + 2X_1 g_{23} \in I^*$, but $h \notin \langle f^*, g_{12}^*, g_{13}^*, g_{23}^* \rangle$ (note that the latter is equal to $\langle f, g_{12}, g_{13}, g_{23} \rangle$ in $k\langle X \rangle[t]$).

Remark In the above examples (iv) and (v) there has been nothing about $G(A)$. However, it will be clear from Theorem 3.7 below that generally the defining relations of $G(A)$ cannot be obtained by simply taking the highest degree homogeneous part of the defining relations of A.

Nevertheless, based on the foregoing relationships in $(*)$, the result $(\bullet)$ recalled above still gives us the light, i.e., we may ask

Question If the defining relations of A form a Gröbner basis, what will happen to the defining relations of $G(A)$ and $\widetilde{A}$?

As a preliminary result, in order to answer the above question, we show that if $\{f_j\}_{j \in J}$ is a standard basis of I in the sense of (e.g., [Gol]), then the defining relations of $\widetilde{A}$ and $G(A)$ can be completely determined.

To see why the standard basis is the first choice in our discussion, we first strengthen previous $(*)$ as follows.

For any $f \in k\langle X \rangle$ we denote by $\mathbf{LH}(f)$ the highest degree homogeneous part of f, i.e., if $f = F_0 + F_1 + \cdots + F_p$ with $F_i \in k\langle X \rangle_i$, then $\mathbf{LH}(f) = F_p$. If I is an ideal of $k\langle X \rangle$, we denote by $\langle \mathbf{LH}(I) \rangle$ the graded ideal generated by $\{\mathbf{LH}(f) \mid f \in I\}$ in $k\langle X \rangle$.

3.1. Proposition Let $A = k\langle X \rangle / I$ and FA the standard filtration on A. Then $G(A) \cong k\langle X \rangle / \langle \mathbf{LH}(I) \rangle$.

Proof We know that $G(A) \cong k\langle X \rangle[t]/(\langle t \rangle + I^*)$. To prove the theorem, we first recall that if $f = F_0 + F_1 + \cdots + F_p \in k\langle X \rangle$, then $\mathbf{LH}(f) = F_p$ and $f^* = \mathbf{LH}(f) + t F_{p-1} + \cdots$. Hence the inclusion map $k\langle X \rangle \hookrightarrow k\langle X \rangle[t]$ yields the inclusion $\langle \mathbf{LH}(f) \rangle \subset \langle t \rangle + I^*$. This, in turn, yields a graded ring homomorphism

$$\frac{k\langle X \rangle}{\langle \mathbf{LH}(f) \rangle} \xrightarrow{\;\varphi\;} \frac{k\langle X \rangle [t]}{(\langle t \rangle + I^*)}$$

$$g + \langle \mathbf{LH}(f) \rangle \quad \mapsto \quad g + (\langle t \rangle + I^*)$$

Obviously, φ is surjective. On the other hand, each element $F \in k\langle X \rangle [t]$ has a unique presentation $F = F_0 + F'$ where $F_0 \in k\langle X \rangle$, $F' \in \langle t \rangle$. Moreover, from CH.I Lemma 4.2 we know that each homogeneous element in I^* is of the form $t^r f^*$ for some $f \in I$. If $f = F_p + F_{p-1} + \cdots + F_0$ with $\mathbf{LH}(f) = F_p$, then $f^* = \mathbf{LH}(f) + t F_{p-1} + \cdots + t^p F_0$. Therefore, each element of $\langle t \rangle + I^*$ can be written as a sum $u + v$, where $u \in \langle \mathbf{LH}(f) \rangle$, $v \in \langle t \rangle$. Thus we can define a ring homomorphism

$$\frac{k\langle X \rangle [t]}{(\langle t \rangle + I^*)} \xrightarrow{\;\psi\;} \frac{k\langle X \rangle}{\langle \mathbf{LH}(I) \rangle}$$

$$F + (\langle t \rangle + I^*) \quad \mapsto \quad F_0 + \langle \mathbf{LH}(I) \rangle$$

Since $\psi \circ \varphi = 1$, it follows that φ is also injective and hence an isomorphism. (Indeed, ψ is the ring homomorphism induced by the canonical homomorphism $k\langle X \rangle [t] \to k\langle X \rangle$ which sends t to 0.) $\square$

Suppose $I = \langle f_j \rangle_{j \in J}$. From the above proposition we certainly expect that $\langle \mathbf{LH}(I) \rangle = \langle \mathbf{LH}(f_j) \rangle_{j \in J}$. This leads to the use of standard bases.

3.2. Definition The set $\{f_j\}_{j \in J}$ is called a *standard basis* of I if each element $f \in I$ with $p = d(f)$, where $f = F_0 + F_1 + \cdots + F_p$ with $F_i \in k\langle X \rangle_i$ and $F_p \neq 0$ (see CH.I §3), has a presentation as a finite sum $f = \sum_j g_j f_j h_j$, where $g_j, h_j \in k\langle X \rangle$, and $d(g_j) + d(f_j) + d(h_j) \leq p$ for all j.

Let I be a graded ideal of $k\langle X \rangle$, i.e., $I = \oplus_{p \in \mathbb{N}} (I \cap k\langle X \rangle_p)$. If $\{f_j\}_{j \in J}$ is a generating set of I consisting of homogeneous elements, then it is easy to see that $\{f_j\}_{j \in J}$ is a standard basis of I. But generally it is not so easy to check if a generating set of an ideal is a standard basis. We refer to [Gol] for a homological criterion of standard basis.

The first easy but important property of a standard basis is the following

3.3. Lemma If $\{f_j\}_{j \in J}$ is a standard basis of I, then for any $f \in I$, $\mathbf{LH}(f) = \sum \mathbf{LH}(g_j) \mathbf{LH}(f_j) \mathbf{LH}(h_j)$ for some $g_j, h_j \in k\langle X \rangle$, $f_j \in \{f_j\}_{j \in J}$. Indeed, we have the more stronger result: $\{f_j\}_{j \in J}$ is a standard basis if and only if $\langle \mathbf{LH}(f_j) \rangle_{j \in J} = \langle \mathbf{LH}(I) \rangle$.

Proof If $\{f_j\}_{j \in J}$ is a standard basis of I, then by the definition it is easy to see that for any $f \in I$, $\mathbf{LH}(f) = \sum \mathbf{LH}(g_j) \mathbf{LH}(f_j) \mathbf{LH}(h_j)$ for some $g_j, h_j \in k\langle X \rangle$,

$f_j \in \{f_j\}_{j \in J}$. Hence $\langle \mathbf{LH}(f_j) \rangle_{j \in J} = \langle \mathbf{LH}(I) \rangle$.

Conversely, if $\langle \mathbf{LH}(f_j) \rangle_{j \in J} = \langle \mathbf{LH}(I) \rangle$, then for any $f \in I$ with $d(f) = p$, say $f = f_p + f_{p-1} + \cdots$ with $f_i \in k\langle X \rangle_i$, we have $\mathbf{LH}(f) = f_p = \sum g_j \mathbf{LH}(f_j) h_j$ for some $g_j, h_j \in k\langle X \rangle$, $f_j \in \{f_j\}_{j \in J}$, and $d(g_j) + d(\mathbf{LH}(f_j)) + d(h_j) = d(g_j) + d(f_j) + d(h_j) = p$. Now the element $f' = f - \sum g_j f_j h_j \in I$ has $d(f') < p$, we may repeat the above argumentation and after a finite number of steps we will reach a presentation $f = \sum g_j f_j h_j$ where $g_j, h_j \in k\langle X \rangle$, $f_j \in \{f_j\}_{j \in J}$ and $d(g_j) + d(f_j) + d(h_j) \leq p$ for all i. It follows that $\{f_j\}_{j \in J}$ is a standard basis of I. $\qquad \square$

3.4. Proposition With notation as before, if $\{f_j\}_{j \in J}$ is a standard basis of I, then

(i) $G(A)$ has defining relations $\mathbf{LH}(f_j)$, $j \in J$; and moreover

(ii) $\{\mathbf{LH}(f_j)\}_{j \in J}$ is a standard basis of $\langle \mathbf{LH}(I) \rangle$.

Proof This follows immediately from Proposition 3.1 and Lemma 3.3. $\qquad \square$

3.5. Proposition With notation as before, if $\{f_j\}_{j \in J}$ is a standard basis of I, then

(i) I^* is generated by $\{f_i^*\}_{i \in J}$, or in other words, the Rees algebra $\tilde{A}$ of A, viewed as a quotient of $k\langle X \rangle[t]$ by previous $(*)$, has defining relations

$$R_i = tX_i - X_i t, \quad i \in \Lambda$$
$$R_j = f_j^*, \qquad j \in J;$$

and moreover

(ii) $\{f_j^*\}_{j \in J}$ is a standard basis of I^*.

Proof By CH.I Lemma 4.2, each homogeneous element in I^* is of the form $t^r f^*$ for some $f \in I$. Suppose $f = \sum_j h_j f_j g_j$. Since $\{f_j\}_{j \in J}$ is a standard basis of I, it follows from Lemma 3.3 and the definition of homogenization of f that

$$d(h_j^*) + d(f_j^*) + d(g_j^*) \leq d(f^*) \text{ and}$$
$$f^* - \sum_j h_j^* f_j^* g_j^* = t^{r_1} m_1^* + t^{r_2} m_2^* + \cdots \text{ with}$$
$$r_j > 0, \ m_j \in I, \text{ and } d(t^{r_j} m_j^*) \leq d(f^*) \text{ for all } m_j.$$

Similarly, for each $m_j^* \in I^*$ where $m_j = \sum_i h_{i_j} f_{i_j} g_{i_j}$, we have

$$d(h_{i_j}^*) + d(f_{i_j}^*) + d(g_{i_j}^*) \leq d(m_j^*) \text{ and}$$
$$m_j^* - \sum_i h_{i_j}^* f_{i_j}^* g_{i_j}^* = t^{r_{1j}} m_{1_j}^* + t^{r_{2j}} m_{2_j}^* + \cdots \text{ with}$$
$$r_{k_j} > 0, \ m_{k_j} \in I, \text{ and } d(t^{r_{k_j}} m_{k_j}^*) \leq d(m_j^*) \text{ for all } m_{k_j}.$$

Since $d(f^*)$ is finite, after a finite number of steps we will reach $f^* \in \langle f_j^* \rangle_{j \in J}$, in particular, $f^* = \sum_j h_j^* f_j^* g_j^*$ with $d(h_j^*) + d(f_j^*) + d(g_j^*) \leq d(f^*)$ for all j. (Note that I is a proper ideal, the final step of the reduction procedure cannot reach an expression like $\sum_l t^l$.) This proves the conclusions of (i) and (ii). $\qquad \square$

Remark One may also obtain Proposition 3.4 from Proposition 3.5. To see this, suppose $I^* = \langle f_j^* \rangle_{j \in J}$. Then since each element F of I^* is of the form $F = \sum_j H_j f_j^* G_j$, $H_i, G_i \in k\langle X \rangle[t]$, we can define the ring homomorphism ψ and accomplish the argumentation as in the proof of Proposition 3.1.

Now we return to use Gröbner bases. Let $k\langle X \rangle$ be the free k-algebra with the k-basis $\mathcal{B}$ consisting of words in the free semigroup $S = \langle X \rangle$, where $X = \{X_i\}_{i \in \Lambda}$. If $(k\langle X \rangle, \mathcal{B}, \succeq_{gr})$ is an admissible system, where $\succeq_{gr}$ is some *graded* monomial ordering on $\mathcal{B}$ (CH.II Definition 1.2), it follows immediately from the definition that

- Any Gröbner basis $\mathcal{G}$ for an ideal I in $k\langle X \rangle$ with respect to $\succeq_{gr}$ is a standard basis of I in the sense of Definition 3.2.

Thus we have reached the following result.

3.6.   Theorem With notation as before, let I be an ideal of $k\langle X \rangle$ and $A = k\langle X \rangle / I$. Consider the standard filtration FA on A. If $\mathcal{G} = \{f_j\}_{j \in J}$ is a Gröbner basis for I with respect to $\succeq_{gr}$, then $G(A)$ has the defining relations

$$R_j = \mathbf{LH}(f_j), \quad j \in J,$$

and $\widetilde{A}$ has the defining relations

$$R_i = tX_i - X_i t, \quad i \in \Lambda,$$
$$R_j = f_j^*, \qquad\quad j \in J.$$

$\square$

Furthermore, let us consider the k-basis

$$\mathcal{B}(t) = \left\{ wt^r \;\middle|\; w \in \mathcal{B}, \; r \geq 0 \right\}$$

for $k\langle X \rangle[t]$. Then the ordering $\succeq_{gr}$ on $\mathcal{B}$ induces a monomial ordering on $\mathcal{B}(t)$, again denoted $\succeq_{gr}$, as follows:

$$w_1 t^{r_1} \succ_{gr} w_2 t^{r_2} \text{ if and only if } w_1 \succ_{gr} w_2, \text{ or } w_1 = w_2 \text{ and } r_1 > r_2.$$

With the definition as abobe, we have $X_j \succ_{gr} t^r$ for all $j \in \Lambda$ and all $r \geq 0$, and we can discuss Gröbner basis in $k\langle X \rangle[t]$ exactly as in $k\langle X \rangle$.

Before mentioning the next theorem we also note that for any nonzero $f \in k\langle X \rangle$, the degree-preserving ordering $\succeq_{gr}$ yields the following equalities:

$$\mathbf{LM}(f) = \mathbf{LM}(\mathbf{LH}(f))$$
$$\mathbf{LM}(f^*) = \mathbf{LM}(f).$$

3.7. **Theorem** With notation as before, let $\mathcal{G} = \{f_j\}_{j \in J} \subset I$ where I is a two-sided ideal of $k\langle X \rangle$. The following are equivalent:

(i) $\mathcal{G}$ is a Gröbner basis of I;

(ii) $\{f_j^*\}_{j \in J}$ is a Gröbner basis of I^* in $k\langle X \rangle[t]$;

(iii) $\{\mathbf{LH}(f_j)\}_{j \in J}$ is a Gröbner basis of the two-sided ideal $\langle \mathbf{LH}(I) \rangle$ in $k\langle X \rangle$.

Proof (i) $\Rightarrow$ (ii). By the above remark, this can be proved exactly as we did in the proof of Proposition 3.5.

(ii) $\Rightarrow$ (i). From CH.I Lemma 4.2 we know that $(f^*)_* = f$ holds for any $f \in I$. So the assertion follows again from the above remark.

(i) $\Leftrightarrow$ (iii). Using the above remark, this can be directly checked. □

For a given finitely generated k-algebra $A = k[a_1, ..., a_n]$, except for $G(A)$ and $\widetilde{A}$ associated to A with respect to the standard filtration FA on A, another graded structure used by several authors in recent years is the so called *homogenized algebra* of A. More precisely, suppose that A is defined subject to the relations

$$F_j(a_1, ..., a_n) = 0, \quad j \in J,$$

where the F_j are elements in the free algebra $k\langle X_1, ..., X_n \rangle$. Then the homogenized (*graded*) algebra of A, usually denoted $H(A)$ in the literature, is defined by the relations:

$$R_i = X_i T - T X_i, \quad 1 \le i \le n,$$
$$R_j = \widetilde{F}_j, \qquad\qquad j \in J,$$

where each $\widetilde{F}_i$ is the homogenization of F_i with respect to T (say from right hand side) in the free k-algebra $k\langle X_1, ..., X_n, T \rangle$ (e.g., if $F = X_1 X_2^2 + X_3$, then $\widetilde{F} = X_1 X_2^2 + X_3 T^2$), or in other words, $H(A) = k\langle X_1, ..., X_n, T \rangle / I$, where I is the two-sided ideal generated by $\{R_i, \widetilde{F}_j \mid 1 \le i \le n, j \in J\}$.

The study of homogenized enveloping algebra was proposed by Smith in [Sm1], and the study of homogenized down-up algebra was proposed by Benkart and Roby in [BR]. The reader is refered to, e.g., [LeS], [LeV] and [Le2] for some representation theory and noncommutative geometry of homogenized enveloping algebras.

From the construction of the homogenized algebra of a given algebra A it is clear that the behavior of $H(A)$ is similar to $\widetilde{A}$, e.g., $H(A)/(1 - \overline{T})H(A) \cong A$, where $\overline{T}$ is the image of T in $H(A)$, and the relations between $\widetilde{A}$ and $H(A)$, $G(A)$ and $H(A)/\overline{T}H(A)$ are given by the following exact sequences:

$$H(A) \to \widetilde{A} \to 0$$
$$H(A)/\overline{T}H(A) \to G(A) \to 0$$

Furthermore, by ([Li1], [LVO4]) there are two basic facts:

a. The lifting (to $\widetilde{A}$) work can be done through $G(A) = \widetilde{A}/X\widetilde{A}$, because X is contained in the graded Jacobson radical of $\widetilde{A}$ (be aware that $\widetilde{A}$ may have zero Jacobson radical); and the lifting (to A) work can be done through $G(A)$ because A is complete with respect to its filtration topology.

b. The dehomogenizing (to A) work can be done through $\widetilde{A}$, because there is a nice relation between the graded module category $\widetilde{A}$-gr and the filtered module category A-filt.

It is easy to see that the similar facts hold for $H(A)$ and $H(A)/\overline{T}H(A)$.

Note that the generators of both $G(A)$ and $H(A)/\overline{T}H(A)$ satisfy simpler relations given by the *highest degree homogeneous part* of the defining relations of A. In order to determine the algebraic properties of $G(A)$ or $H(A)/\overline{T}H(A)$, it remains only to find the defining relations of $G(A)$ or $H(A)/TH(A)$.

Extending the notion of homogenized algebra to arbitrary k-algebra, below we apply the results of previous sections to assert that in many cases $\widetilde{A} \cong H(A)$.

3.8. Corollary With notation as above, the equivalent conditions of Theorem 3.7 are equivalent to one of the following two conditions.

(i) $\{X_iT - TX_i, \ \widetilde{f}_j \mid i \in \Lambda, \ j \in J\}$ forms a Gröbner basis in $k\langle X_i, T\rangle_{i \in \Lambda}$ with respect to $\geq_{grlex}$ such that $X_i >_{grlex} T$, $i \in \Lambda$.

(ii) $\widetilde{A} \cong H(A)$, where $H(A)$ is the homogenized (graded) algebra of $A = k[x_i]_{i \in \Lambda}$.

$\square$

Certainly, Theorem 3.7 and Corollary 3.8 can be applied to general quadric solvable polynomial algebras defined in previous §2, but we leave this to CH.IV §4 for completeness.

Let us finish this section by returning to the examples given in CH.I §5.

3.9. Corollary If A is one of the k-algebras listed in CH.I §5 Example (i)–(vii), then, after setting a suitable ordering on the generators in order to define $\geq_{grlex}$ as in the examples of §2, one may make Proposition 2.2 and Proposition 1.6 hold, and therefore, the defining relations of A form a Gröbner basis in the corresponding free algebra with respect to $\geq_{grlex}$. (Note that in [Mor2] it has been verified that the defining relations of $A_n(k)$, respectively the defining relations of $U(\mathbf{g})$, form a Gröbner basis.). Moreover, one may further use Theorem 3.6– Corollary 3.8 to get the defining relations for the associated graded structures of A, in particular, $G(A)$ and $\widetilde{A}$ are quadratic algebras and most of them are iterated skew polynomial algebras starting with the ground field k. For instance, we mention one of the following consequences:

(i) The associated graded algebra $G(A_n(k))$ of the nth Weyl algebra $A_n(k)$ over a field k has the defining relations

$$X_jX_i - X_iX_j, \quad i \neq j, \ i,j = 1,...,n.$$

Hence $G(A_n(k))$ is isomorphic to the commutative polynomial k-algebra in $2n$

variables. And the Rees algebra $\widetilde{A_n(k)}$ of $A_n(k)$ has the defining relations

$$Y_j Y_i - Y_i Y_j, \; X_j X_i - X_i X_j, \quad 1 \le i < j \le n,$$
$$Y_i T - T Y_i, \; X_i T - T X_i, \qquad 1 \le i \le n,$$
$$Y_j X_i - X_i Y_j - \delta_{ij} T^2, \qquad 1 \le i, j \le n,$$

that make $\widetilde{A_n(k)}$ into an iterated skew polynomial algebra.

(ii) Let $\mathbf{g}$ be a finite$n$- dimensional Lie algebra over a field k and $U(\mathbf{g})$ its enveloping algebra. Then the associated graded algebra $G(U(\mathbf{g}))$ of $U(\mathbf{g})$ has the defining relations

$$X_j X_i - X_i X_j, \quad i \ne j, \, 1 \le i < j \le n.$$

Hence $G(U(\mathbf{g}))$ is isomorphic to the commuttaive polynomial k-algebra in n variables. And the Rees algebra $\widetilde{U(\mathbf{g})}$ of $U(\mathbf{g})$ has the defining relations

$$X_i T - T X_i, \qquad\qquad 1 \le i \le n,$$
$$X_j X_i - X_i X_j - \sum_{\ell=1}^{n} \lambda_{ij\ell} X_\ell T, \quad 1 \le i < j \le n.$$

It follows that the so called homogenized enveloping algebra proposed in [Sm1] and studied in [LeV] and [LeS] is nothing but exactly the Rees algebra of $U(g)$.

4. A Remark on Recognizable Properties of Algebras via Gröbner Bases

For a given k-algebra A and a certain algebraic property P of A, in general we hardly know whether or not P can be algorithmically determined (e.g., see [Ufn1]). However, we have seen from CH.II §4 that a stronger Noetherian property, i.e., the G-Noetherian property of an algebra A may be realized by finding a (left) Dickson system associated with A. Moreover, from previous sections we also have seen that given a k-algebra $A = k\langle X \rangle / I$, where $k\langle X \rangle$ is the free k-algebra over $X = \langle X_i \rangle_{i \in J}$ and I is an ideal of $k\langle X \rangle$, if I has a Gröbner basis, then the structural properties of A, such as the ideal membership and to have a PBW basis, can be algorithmically established. Not only this, indeed, the earlier work of [Berg] has motivated more discovery of recognizable properties of algebras via very noncommutative Gröbner bases. To be convincible, it is the aim of this remark section to introduce the work of [G-IL] which establishes the existence of algorithms for recognizing certain structural properties of finitely generated k-algebras. We refer the reader to [G-IL] for detailed proofs of the statements mentioned in this section and for relevant algorithms written in pseudo-codes.

In what follows, $k\langle X\rangle = k\langle X_1, ..., X_n\rangle$, i.e., the algebra $A = k\langle X\rangle/I$ is finitely generated, and we suppose that the two-sided ideal I is generated by $\mathcal{G} = \{g_1, ..., g_s\}$, i.e., A is *finitely presented*.

4.1. Theorem ([G-IL] Theorem 1) Suppose that $\mathcal{G}$ is a Gröbner basis for I. Then the following properties of A are algorithmically recognizable:

(i) A has a growth of a given type, in particular, A is finite dimensional.

(ii) A is algebraic over k.

(iii) A satisfies the polynomial identity $[X_1, ..., X_r]$ of Lie nilpotency, for a fixed integer r.

$\square$

4.2. Theorem ([G-IL] Theorem 2) Suppose that the generating set $\mathcal{G}$ of I consists of monomials (hence $\mathcal{G}$ is a Gröbner basis for I by CH.II Proposition 3.5). Then the following properties of A are algorithmically recognizable:

(i) A has no nilpotent elements.

(ii) A is semisimple (in the sense of Jacobson).

(iii) A is prime.

(iv) A is semiprime.

$\square$

Note that the algebra A appearing in Theorem 4.2 above has been called a *monomial algebra* in [G-IL]. We also refer to [G-I1–3] and [Ufn1] for more discussion in algorithmically recognizing some other structural properties of finitely presented algebras, such as Jacobson radical, Hilbert and Poincaré series, global (homological) dimension, and noetherianity, etc.

In CH.VIII, we will see that the results obtained in §1–§3 of this chapter and CH.IV §4 enable us to show that every quadric solvable polynomial algebra has finite global dimension and K_0-group $\mathbb{Z}$.

CHAPTER IV
Filtered-Graded Transfer of Gröbner Bases

Let $A = k[a_i]_{i \in \Lambda}$ be a k-algebra generated by $\{a_i\}_{i \in \Lambda}$ over the field k. Let FA be the standard filtration on A, $G(A) = \oplus_{p \in \mathbb{N}} G(A)_p$ with $G(A)_p = F_p A / F_{p-1} A$, the associated graded algebra of A, and $\widetilde{A} = \oplus_{p \in \mathbb{N}} \widetilde{A}_p$ with $\widetilde{A}_p = F_p A$, the Rees algebra of A in the sense of CH.I §3. Then by CH.I Proposition 3.16,

a. $G(A) = k[\sigma(a_i)]_{i \in \Lambda}$ where each $\sigma(a_i)$ is the image of a_i in $F_1 A / F_0 A = G(A)_1$, and

b. $\widetilde{A} = k[X, \widetilde{a}_i]_{i \in \Lambda}$ where X is the canonical element of degree 1 represented by 1 in $\widetilde{A}_1 = F_1 A$ and each $\widetilde{a}_i$ is the homogeneous element of degree 1 represented by a_i in $\widetilde{A}_1 = F_1 A$.

Suppose that $A \cong k\langle X \rangle / \langle f_j \rangle_{j \in J}$, where $k\langle X \rangle$ is the free k-algebra on $X = \{X_i\}_{i \in \Lambda}$ and $\langle f_j \rangle_{j \in J}$ is the two-sided ideal of $k\langle X \rangle$ generated by $\{f_j\}_{j \in J}$. Let $\mathcal{B}$ be the k-basis of $k\langle X \rangle$ consisting of all words of the free semigroup $S = \langle X \rangle$. If $\{f_j\}_{j \in J}$ forms a Gröbner basis with respect to a graded monomial ordering $\succeq_{gr}$ on $\mathcal{B}$, then CH.III Theorem 3.6 entails that, with respect to the standard filtration of FA on A,

c. $G(A) \cong k\langle X \rangle / \langle \mathbf{LH}(f_j) \rangle_{j \in J}$ and $\widetilde{A} \cong k\langle X \rangle [t] / \langle f_j^* \rangle_{j \in J}$, where each $\mathbf{LH}(f_j)$ is the highest degree homogeneous part of f_j in $k\langle X \rangle$, and each f_j^* is the homogenization of f_j in the polynomial ring $k\langle X \rangle [t]$ with respect to the commuting variable t.

However, if we consider the standard filtration (or the grading filtration) on

$k\langle X\rangle$, it follows from CH.I Proposition 3.18 that $G(k\langle X\rangle) \cong k\langle X\rangle$ and $\widetilde{k\langle X\rangle} \cong k\langle X\rangle[t]$. In other words, an application of CH.III Theorem 3.6 to $k\langle X\rangle$ indeed yields the following equivalence:

d. $\{f_j\}_{j\in J}$ is a Gröbner basis in $k\langle X\rangle$.

e. $\{\mathbf{LH}(f_j)\}_{j\in J}$ is a Gröbner basis in $G(k\langle X\rangle)$.

f. $\{f_j^*\}_{j\in J}$ is a Gröbner basis in $\widetilde{k\langle X\rangle}$.

Noticing the the above facts and the interpretation given in CH.III §3, in this chapter we develop a general philosophy of the filtered-graded transfer of Gröbner bases so that, on one hand, certain Gröbner basis theory may be valid for more algebras, and on the other hand, certain computations using noncommutative Gröbner bases may possibly be carried out first at a *feasible graded level* and then be lifted (or dehomogenized) back to *ungraded level* (as illustrated in later CH.V–CH.VIII).

To be convenient, we fix the convention of this chapter once and for all.
Let $A = k[a_i]_{i\in\Lambda}$ be a k-algebra as above, $\mathcal{B}$ a fixed k-basis of A consisting of monomials, that is,

$$\mathcal{B} = \left\{ w = a_{i_1}^{\alpha_1} \cdots a_{i_n}^{\alpha_n} \;\middle|\; n \geq 1,\ \alpha_i \in I\!N \right\},$$

and FA the standard filtration on A determined by $\mathcal{B}$:

$$F_pA = \left\{ \sum c_i w_i \;\middle|\; c_i \in k,\ w_i \in \mathcal{B},\ d(w_i) \leq p \right\}, \quad p \in I\!N,$$

where $d(w)$ denote the *degree* of w (see CH.II §1, e.g., if $w = a_{i_1}^{\alpha_1} \cdots a_{i_n}^{\alpha_n}$ then $d(w) = |\alpha| = \alpha_1 + \cdots + \alpha_n$). We call this filtration the $\mathcal{B}$-*standard filtration on* A. Moreover, assume that $\mathcal{B}$ is a *strictly filtered basis* in the sense that

- $\mathcal{B} = \bigcup_{p\geq 0} \mathcal{B}_p$ with $\mathcal{B}_p = \{w \in \mathcal{B} \mid d(w) \leq p\}$, and if $d(w) = p$ then $w \notin \mathcal{B}_{p-1}$.

Let $A = k[a_1, ..., a_n]$ be a finitely generated k-algebra. If the set of all standard monomials

$$\mathcal{B} = \left\{ a_1^{\alpha_1} \cdots a_n^{\alpha_n} \;\middle|\; (\alpha_1, ..., \alpha_n) \in I\!N^n \right\}$$

forms a k-basis (i.e., a PBW basis in the sense of CH.III §1), then, the definition of $\mathcal{B}$-standard filtration FA on A shows clearly that $\mathcal{B}$ is strictly filtered. So the assumption (•) is indeed quite mild.

1. Filtered-Graded Transfer of (Left) Admissible Systems

Let A, $\mathcal{B}$ and FA be as assumed in the beginning of this chapter. Let $G(A) = \oplus_{p\geq 0} G(A)_p$ and $\widetilde{A} = \oplus_{p\geq 0} \widetilde{A}_p$ be the associated graded algebra   and the Rees algebra of A, respectively. With notation as in CH.I §3, let v be the order function on A determined by FA. If $f \in A$ with $v(f) = p$ (i.e., $f \in F_pA - F_{p-1}A$), then the homogeneous element represented by f in $G(A)_p$, respectively in $\widetilde{A}_p$ is denoted by $\sigma(f)$, respectively by $\widetilde{f}$.

In this section we discuss the filtered-graded transfer of (left) admissible systems associated with A, $G(A)$, and $\widetilde{A}$.

First note some basic facts about a strictly filtered k-basis $\mathcal{B}$ defined by the foregoing ($\bullet$).

1.1. Lemma (i) For each $w \in \mathcal{B}$, $d(w) = v(w)$.
(ii) For each $p \in I\!N$, $\mathcal{B}_p$ is a k-basis of F_pA.
(iii) For $w, u \in \mathcal{B}$, $\sigma(u) = \sigma(w)$ if and only if $u = w$. □

Next we establish the filtered-graded transfer of k-bases between A, $G(A)$, and $\widetilde{A}$, respectively.

1.2. Lemma For any $w \in \mathcal{B}$, if $w = a_{i_1}^{\alpha_1} \cdots a_{i_n}^{\alpha_n}$ with $d(w) = p$, then $\sigma(w) = \sigma(a_{i_1})^{\alpha_1} \cdots \sigma(a_{i_1})^{\alpha_n} \in G(A)_p$ and $\widetilde{w} = \widetilde{a}_{i_1}^{\alpha_1} \cdots \widetilde{a}_{i_n}^{\alpha_n} \in \widetilde{A}_p$.

Proof Since $w \neq 0$ (element of the k-basis), we conclude by the assumption ($\bullet$) on $\mathcal{B}$ that $w = a_{i_1}^{\alpha_1} \cdots a_{i_n}^{\alpha_n} \in F_pA - F_{p-1}A$. Hence, $\sigma(a_{i_1})^{\alpha_1} \cdots \sigma(a_{i_1})^{\alpha_n} \neq 0$ and it follows from CH.I Lemma 3.11 that $\sigma(w) = \sigma(a_{i_1})^{\alpha_1} \cdots \sigma(a_{i_1})^{\alpha_n} \in G(A)_p$ and $\widetilde{w} = \widetilde{a}_{i_1}^{\alpha_1} \cdots \widetilde{a}_{i_n}^{\alpha_n} \in \widetilde{A}_p$. □

1.3. Proposition Let A, $\mathcal{B}$, and FA be as fixed. Then the following are equivalent.
(i) $\mathcal{B}$ is a strictly filtered k-basis for A.
(ii) $\sigma(\mathcal{B}_p) = \{\sigma(w) \mid w \in \mathcal{B}_p, \ d(w) = p\}$ forms a k-basis of $G(A)_p$ for each $p \in I\!N$. Hence $\sigma(\mathcal{B}) = \left\{\sigma(w) \mid w \in \mathcal{B}\right\}$ forms a k-basis for $G(A)$.
(iii) $\widetilde{\mathcal{B}}_p = \{\widetilde{w}X^{p-s} \mid w \in \mathcal{B}_p, \ d(w) = s \leq p\}$ forms a k-basis of $\widetilde{A}_p$ for each $p \in I\!N$. Hence $\widetilde{\mathcal{B}} = \{\widetilde{w}X^h \mid w \in \mathcal{B}, \ h \in I\!N\}$ forms a k-basis for $\widetilde{A}$, where X is the canonical element of degree 1 in $\widetilde{A}$ (CH.I §3).

Proof (i) $\Leftrightarrow$ (ii) Let $f = \sum c_i w_i$ be an element of A with $c_i \in k$, $w_i \in \mathcal{B}$. If $f \in F_pA - F_{p-1}A$, then since FA is the $\mathcal{B}$-standard filtration, we have $w_i \in \mathcal{B}_p$ and $\sigma(f) = \sum_{d(w_i)=p} c_i \sigma(w_i)$. Thus, $G(A)_p$ is, as a k-space, spanned by $\sigma(\mathcal{B}_p)$. If $\sum_i \lambda_i \sigma(w_i) = 0$ for $\lambda_i \in k$ and $\sigma(w_i) \in \sigma(\mathcal{B}_p)$, then $\sum_i \lambda_i w_i \in F_{p-1}A$. But this implies $\lambda_i = 0$ for all i as $\mathcal{B}$ is a strictly filtered basis by (i). Since FA is separated and $d(w) = v(w)$ for $w \in \mathcal{B}$ by Lemma 1.1, the implication (ii) $\Rightarrow$ (i) may be directly verified.
(i) $\Leftrightarrow$ (iii) Note that $\sum_{w_i \in \mathcal{B}_p} \lambda_i \widetilde{w}_i X^{p-s_i} = 0$ implies $\sum_i \lambda_i w_i = 0$ and $\lambda_i = 0$ by CH.I Proposition 3.8. It is sufficient to show that $\widetilde{A}_p$ is spanned by $\widetilde{\mathcal{B}}_p$. By

CH.I Lemma 3.11, every homogeneous element F in $\widetilde{A}_p$ is of the form $\widetilde{f}X^{p-s}$, where $s = v(f) \leq p$, i.e., $f \in F_sA - F_{s-1}A$. Suppose that $f = \sum_i c_iw_i$ with $c_i \in k$ and $w_i \in \mathcal{B}$. Since FA is the $\mathcal{B}$-standard filtration, $w_i \in F_sA \subseteq F_pA$ and $\widetilde{f} = \sum_i c_i\widetilde{w_i}X^{s-q_i}$ where $q_i = d(w_i) = v(w_i) = q_i \leq s$. Hence, $F = \widetilde{f}X^{p-s} = \sum_i c_i\widetilde{w_i}X^{p-q_i}$ with $q_i \leq s \leq p$, as desired. Similarly the implication (iii) $\Rightarrow$ (i) may be verified. $\square$

Now, let $(A, \mathcal{B}, \succeq_{gr})$ be a (left) admissible system with $\succeq_{gr}$ a graded monomial ordering on $\mathcal{B}$ in the sense of CH.II Definition 1.2(iii). With notation as in Proposition 1.3, define on $\sigma(\mathcal{B})$, respectively on $\widetilde{\mathcal{B}}$, the ordering

$$(*) \quad \begin{cases} \sigma(u) \succ_{gr} \sigma(w) \text{ if and only if } u \succ_{gr} w, \\ \text{respectively} \\ \widetilde{u}X^h \succ_{gr} \widetilde{w}X^\ell \text{ if and only if } u \succ_{gr} w \text{ or } u = w \text{ and } h > \ell. \end{cases}$$

From the definition $(*)$ above we observe that

  a. $\widetilde{w} \succ_{gr} X^p$ for all $w \in \mathcal{B}$ with $w \neq 1$ and all $p \geq 0$.
  b. $\succeq_{gr}$ is a well-ordering on $\sigma(\mathcal{B})$, respectively on $\widetilde{\mathcal{B}}$.

The next lemma shows that $\succeq_{gr}$ is *compatible* with the $\mathcal{B}$-standard filtration FA on A.

1.4. Lemma Let $(A, \mathcal{B}, \succeq_{gr})$ be as above. Then the following holds.
(i) For $f \in A$, $f \in F_pA$ if and only if $d(\mathbf{LM}(f)) = v(f) \leq p$, and $f \in F_pA - F_{p-1}A$ if and only if $d(\mathbf{LM}(f)) = p$; for $f, g \in A$, $\mathbf{LM}(f) \succeq_{gr} \mathbf{LM}(g)$ implies $d(\mathbf{LM}(f)) = v(f) \geq v(g) = d(\mathbf{LM}(g))$.
(ii) For each $w \in \mathcal{B}$, $\sigma(w)$, respectively $\widetilde{w}$, is a monomial of $G(A)$, respectively a monomial of $\widetilde{A}$, and moreover $d(w) = d(\sigma(w)) = d(\widetilde{w})$.
(iii) With the definition as in $(*)$ above, for $f \in A$, $\sigma(\mathbf{LM}(f)) = \mathbf{LM}(\sigma(f))$, $\widetilde{\mathbf{LM}(f)} = \mathbf{LM}(\widetilde{f})$, and $\mathbf{LM}(X^s\widetilde{f}) = X^s\mathbf{LM}(\widetilde{f})$ for all $s \geq 0$.

Proof (i) Since $\mathcal{B}$ is strictly filtered and FA is the $\mathcal{B}$-standard filtration, this is clear by Lemma 1.1.
(ii) This follows from Lemma 1.2.
(iii) If $f \in F_pA - F_{p-1}A$, we may write $f = \sum_{i=1}^n \lambda_iw_i$, where $\lambda_i \in k - \{0\}$, such that $w_i \in \mathcal{B}_p$ and $w_1 \succ_{gr} w_2 \succ_{gr} \cdots \succ_{gr} w_n$. Then $\mathbf{LM}(f) = w_1$ and $d(\mathbf{LM}(f)) = v(f) = d(w_1) = p$ by (i). Thus, $\sigma(f) = \sum_{d(w_i)=p} \lambda_i\sigma(w_i)$ and it follows from the definition $(*)$ that $\mathbf{LM}(\sigma(f)) = \sigma(w_1) = \sigma(\mathbf{LM}(f))$. With the same f, we have $\widetilde{f} = \lambda_1\widetilde{w_1} + \sum_{i=2}^n \lambda_i\widetilde{w_i}X^{p-s_i}$, where $s_i = d(w_i) = v(w_i) < p$, and it follows from the definition $(*)$ that $\mathbf{LM}(\widetilde{f}) = \widetilde{w_1} = \widetilde{\mathbf{LM}(f)}$, and $\mathbf{LM}(X^s\widetilde{f}) = X^s\mathbf{LM}(\widetilde{f})$ for all $s \geq 0$. This finishes the proof. $\square$

We are ready to mention the main result of this section.

1.5. Theorem Let FA be the $\mathcal{B}$-standard filtration on A. The following statements hold.

(i) If $(A, \mathcal{B}, \succeq_{gr})$ is a (left) admissible system, where $\succeq_{gr}$ is some graded monomial ordering on $\mathcal{B}$, then $(G(A), \sigma(\mathcal{B}), \succeq_{gr})$ is a (left) admissible system in which $\succeq_{gr}$ is defined as in the foregoing (∗).

(ii) If $(A, \mathcal{B}, \succeq_{gr})$ is a (left) admissible system, where $\succeq_{gr}$ is some graded monomial ordering on $\mathcal{B}$, then $(\widetilde{A}, \widetilde{\mathcal{B}}, \succeq_{gr})$ is a (left) admissible system in which $\succeq_{gr}$ is defined as in the foregoing (∗).

Proof (i) Let $(A, \mathcal{B}, \succeq_{gr})$ be as assumed. By Proposition 1.3 and the observations mentioned below the definition (∗), we only need to verify CH.II Definition 1.2 (**MO1**)–(**MO2**) for the elements of $\sigma(\mathcal{B})$. using $\succeq_{gr}$ as defined in the foregoing (∗), if $\sigma(w)$, $\sigma(u)$, $\sigma(v)$, $\sigma(s) \in \sigma(\mathcal{B})$ with $\sigma(u) \succ_{gr} \sigma(w)$, and if $\mathbf{LM}(\sigma(v)\sigma(w)\sigma(s)) \neq 0$ and $\mathbf{LM}(\sigma(v)\sigma(u)\sigma(s)) \neq 0$, then

$$
(1) \quad \begin{cases} u \succ_{gr} w, \\ \sigma(vws) = \sigma(v)\sigma(w)\sigma(s) \neq 0, \\ \sigma(vus) = \sigma(v)\sigma(u)\sigma(s) \neq 0, \\ \mathbf{LM}(vws) \neq 0, \quad \mathbf{LM}(vus) \neq 0. \end{cases} \quad \text{(CH.I Lemma 3.11)}
$$

Hence, $\mathbf{LM}(vus) \succ_{gr} \mathbf{LM}(vws)$ and this yields $\sigma(\mathbf{LM}(vus)) \succ_{gr} \sigma(\mathbf{LM}(vws))$. By Lemma 1.4 and (1) above, we have

$$
\sigma(\mathbf{LM}(vus)) = \mathbf{LM}(\sigma(vus)) = \mathbf{LM}(\sigma(v)\sigma(u)\sigma(s)),
$$
$$
\sigma(\mathbf{LM}(vws)) = \mathbf{LM}(\sigma(vws)) = \mathbf{LM}(\sigma(v)\sigma(w)\sigma(s)).
$$

Therefore, $\mathbf{LM}(\sigma(v)\sigma(u)\sigma(s)) \succ_{gr} \mathbf{LM}(\sigma(v)\sigma(w)\sigma(s))$. This verifies (**MO1**). If $\sigma(u) = \mathbf{LM}(\sigma(v)\sigma(w)\sigma(s))$ with $\sigma(v) \neq 1$ or $\sigma(s) \neq 1$, then by CH.I Lemma 3.11 and the foregoing Lemma 1.4, we obtain that $\sigma(u) = \mathbf{LM}(\sigma(vws)) = \sigma(\mathbf{LM}(vus))$ and consequently $u = \mathbf{LM}(vws)$ (Lemma 1.1). But this implies $u \succ_{gr} w$. Hence, $\sigma(u) \succ_{gr} \sigma(w)$ and this verifies (**MO2**) as well.

(ii) Again, by Proposition 1.3 we only need to verify CH.II Definition 1.2 (**MO1**)–(**MO2**). And by Lemma 1.4(iii), we only need to deal with the elements in $\widetilde{\mathcal{B}}$ which are not of the form $X^n \widetilde{u}$ with $n > 0$. Using $\succeq_{gr}$ as defined in the foregoing (∗), if $\widetilde{w}$, $\widetilde{u}$, $\widetilde{v}$, $\widetilde{s} \in \widetilde{\mathcal{B}}$ with $\widetilde{u} \succ_{gr} \widetilde{w}$, and if $\mathbf{LM}(\widetilde{v}\widetilde{w}\widetilde{s}) \neq 0$, $\mathbf{LM}(\widetilde{v}\widetilde{u}\widetilde{s}) \neq 0$, then

$$
(2) \quad \begin{cases} u \succ_{gr} w, \\ X^n \widetilde{vws} = \widetilde{v}\widetilde{w}\widetilde{s} \neq 0 \text{ for some } n \geq 0, \\ X^m \widetilde{vus} = \widetilde{v}\widetilde{u}\widetilde{s} \neq 0 \text{ for some } m \geq 0, \\ \mathbf{LM}(vws) \neq 0, \quad \mathbf{LM}(vus) \neq 0. \end{cases} \quad \text{(CH.I Lemma 3.11)}
$$

Hence $\mathbf{LM}(vus) \succ_{gr} \mathbf{LM}(vws)$ and consequently $\widetilde{(\mathbf{LM}(vus))} \succ_{gr} \widetilde{(\mathbf{LM}(vws))}$.

By Lemma 1.4 and (2) above we obtain

$$X^m(\mathbf{LM}(vus))\widetilde{} = \mathbf{LM}(X^m\widetilde{vus}) = \mathbf{LM}(\widetilde{v}\widetilde{u}\widetilde{s}),$$
$$X^n(\mathbf{LM}(vws))\widetilde{} = \mathbf{LM}(X^n\widetilde{vws}) = \mathbf{LM}(\widetilde{v}\widetilde{w}\widetilde{s}).$$

Hence, $\mathbf{LM}(\widetilde{v}\widetilde{u}\widetilde{s}) \succ_{gr} \mathbf{LM}(\widetilde{v}\widetilde{w}\widetilde{s})$, i.e., (**MO1**) is verified.

If $\widetilde{u} = \mathbf{LM}(\widetilde{v}\widetilde{w}\widetilde{s})$ with $\widetilde{v} \neq 1$ or $\widetilde{s} \neq 1$, then by CH.I Lemma 3.11 and the foregoing Lemma 1.4 we have $\widetilde{u} = \mathbf{LM}(\widetilde{v}\widetilde{w}\widetilde{s}) = \mathbf{LM}(\widetilde{vws}X^n) = X^n(\mathbf{LM}(vws))\widetilde{}$ for some $n \neq 0$. Hence $u = \mathbf{LM}(vws)$. But this implies $u \succ_{gr} w$ and consequently $\widetilde{u} \succ_{gr} \widetilde{w}$. This verifies (**MO2**).   □

We finish this section by the following theorem which may be viewed as the converse of Theorem 1.5 but under a stronger condition.

1.6. Theorem Let FA be the $\mathcal{B}$-standard filtration on A. Suppose that $G(A)$ is a domain. The following statements hold.

(i) If $(G(A), \sigma(\mathcal{B}), \succeq_{gr})$ is a (left) admissible system, where $\succeq_{gr}$ is some graded monomial ordering on $\sigma(\mathcal{B})$, then $(A, \mathcal{B}, \succeq_{gr})$ is a (left) admissible system in which $\succeq_{gr}$ is defined as

$$u \succ_{gr} w \text{ if and only if } \sigma(u) \succ_{gr} \sigma(w), \ u, w \in \mathcal{B}.$$

(Note that by Lemma 1.1, $\sigma(u) = \sigma(w)$ implies $u = w$.)

(ii) If $(\widetilde{A}, \widetilde{\mathcal{B}}, \succeq_{gr})$ is a (left) admissible system, where $\succeq_{gr}$ is some graded monomial ordering on $\widetilde{\mathcal{B}}$ such that $\widetilde{w} \succ_{gr} X^s$ for all $s \geq 0$, then $(A, \mathcal{B}, \succeq_{gr})$ is a (left) admissible system in which $\succeq_{gr}$ is defined as

$$u \succeq_{gr} w \text{ if and only if } \widetilde{u} \succeq_{gr} \widetilde{w}, \ u, w \in \mathcal{B}.$$

(Note that by the definition of $\widetilde{u}$, $\widetilde{u} = \widetilde{w}$ implies $u = w$.)

Proof (i) By Proposition 1.3 we only need to verify CH.II Definition 1.2 (**MO1**)–(**MO2**) for the elements of $\mathcal{B}$. Let $\succeq_{gr}$ be as defined in the first part of the theorem. If $u, w, v, s \in \mathcal{B}$ with $u \succ_{gr} w$, and if $\mathbf{LM}(vus) \neq 0$, $\mathbf{LM}(vws) \neq 0$, then since $G(A)$ is a domain, we have by Lemma 1.4 that

$$(3) \quad \begin{cases} \sigma(u) \succ_{gr} \sigma(w), \\ \sigma(\mathbf{LM}(vus)) = \mathbf{LM}(\sigma(vus)) = \mathbf{LM}(\sigma(v)\sigma(u)\sigma(s)) \neq 0, \\ \sigma(\mathbf{LM}(vws)) = \mathbf{LM}(\sigma(vws)) = \mathbf{LM}(\sigma(v)\sigma(w)\sigma(s)) \neq 0, \\ \mathbf{LM}(\sigma(v)\sigma(u)\sigma(s)) \succ_{gr} \mathbf{LM}(\sigma(v)\sigma(w)\sigma(s)). \end{cases}$$

Hence $\sigma(\mathbf{LM}(vus)) \succ_{gr} \sigma(\mathbf{LM}(vws))$ and consequently $\mathbf{LM}(vus) \succ_{gr} \mathbf{LM}(vws)$. This verifies (**MO1**).

If $u = \mathbf{LM}(vws)$ with $v \neq 1$ or $s \neq 1$, then since $G(A)$ is a domain, $\sigma(u) = \sigma(\mathbf{LM}(vws)) = \mathbf{LM}(\sigma(vws)) = \mathbf{LM}(\sigma(v)\sigma(w)\sigma(s))$ by Lemma 1.4. This implies $\sigma(u) \succ_{gr} \sigma(w)$ and consequently $u \succ_{gr} w$, i.e., (**MO2**) is verified as well.

(ii) By Proposition 1.3 we only need to verify CH.II Definition 1.2 (**MO1**)–(**MO2**) for the elements of $\mathcal{B}$. Let $\succeq_{gr}$ be as defined in the second part of the theorem. If $u, w, v, s \in \mathcal{B}$ with $u \succ_{gr} w$, and if $\mathbf{LM}(vus) \neq 0$, $\mathbf{LM}(vws) \neq 0$, then since $G(A)$ is a domain, we have by Lemma 1.4 that

$$
(4) \quad
\begin{cases}
\widetilde{u} \succ_{gr} \widetilde{w}, \\
(\mathbf{LM}(vus))\widetilde{} = \mathbf{LM}(\widetilde{vus}) = \mathbf{LM}(\widetilde{v}\widetilde{u}\widetilde{s}) \neq 0, \\
(\mathbf{LM}(vws))\widetilde{} = \mathbf{LM}(\widetilde{vws}) = \mathbf{LM}(\widetilde{v}\widetilde{w}\widetilde{s}) \neq 0, \\
\mathbf{LM}(\widetilde{v}\widetilde{u}\widetilde{s}) \succ_{gr} \mathbf{LM}(\widetilde{v}\widetilde{w}\widetilde{s})
\end{cases}
$$

Hence $(\mathbf{LM}(vus))\widetilde{} \succ_{gr} (\mathbf{LM}(vws))\widetilde{}$ and consequently $\mathbf{LM}(vus) \succ_{gr} \mathbf{LM}(vws)$. This verifies (**MO1**).

If $u = \mathbf{LM}(vws)$ with $v \neq 1$ or $s \neq 1$, then since $G(A)$ is a domain, $\widetilde{u} = (\mathbf{LM}(vws))\widetilde{} = \mathbf{LM}(\widetilde{vws}) = \mathbf{LM}(\widetilde{v}\widetilde{w}\widetilde{s})$ by Lemma 1.4. This implies $\widetilde{u} \succ_{gr} \widetilde{w}$ and consequently $u \succ_{gr} w$, i.e., (**MO2**) is verified as well. This completes the proof. □

2. Filtered-Graded Transfer of (Left) Gröbner Bases

Let $A = k[a_i]_{i \in \Lambda}$, $\mathcal{B}$ a strictly filtered k-basis of A, and FA the $\mathcal{B}$-standard filtration on A. If L is a (left) ideal of A with the filtration FL induced by FA on L, then there are the associated (left) graded ideals $G(L) \subset G(A)$ and $\widetilde{L} \subset \widetilde{A}$ (CH.I §3). In this section we discuss the filtered-graded transfer of Gröbner bases between L, $G(L)$, and $\widetilde{L}$.

2.1. Theorem With notation as above, suppose that $(A, \mathcal{B}, \succeq_{gr})$ is a (left) admissible system where $\succeq_{gr}$ is some graded monomial ordering on $\mathcal{B}$, and that $G(A)$ is a domain. Let I be a left ideal or a two-sided ideal of A. The following statements hold.

(i) If $\mathcal{G} = \{g_i\}_{i \in J}$ is a (left) Gröbner basis for I, then $\sigma(\mathcal{G}) = \{\sigma(g_i) \mid g_i \in \mathcal{G}\}$ is a (left) Gröbner bsis for $G(I)$, with respect to the (left) admissible system $(G(A), \sigma(\mathcal{B}), \succeq_{gr})$ as determined in Theorem 1.5.

(ii) If $\{G_i\}_{i \in J}$ is a (left) Gröbner basis of $G(I)$ consisting of homogeneous elements, with respect to the (left) admissible system $(G(A), \sigma(\mathcal{B}), \succeq_{gr})$ as determined in Theorem 1.5, and if g_i is choosen to be such that $\sigma(g_i) = G_i$, $i \in J$, then $\mathcal{G} = \{g_i\}_{i \in J}$ is a (left) Gröbner basis for I.

Proof We prove the theorem for a two-sided ideal I of A (a similar argumentation works for a left ideal).

(i) We show that every nonzero homogeneous element $F \in G(I)$ has a Gröbner presentation by $\sigma(\mathcal{G})$ in the sense of CH.II Definition 3.2. Suppose $F \in G(A)_p$. Then $F = \sigma(f)$ for some $f \in F_pA \cap I - F_{p-1}A$, and f has a Gröbner presentation by $\mathcal{G}$, say $f = \sum_{i=1}^{s} \lambda_i u_i g_i v_i$, where $\lambda_i \in k-\{0\}$, $u_i, v_i \in \mathcal{B}$, $g_i \in \mathcal{G}$, with the property that

$$\mathbf{LM}(f) \succeq_{gr} \mathbf{LM}(u_i g_i v_i) \text{ whenever } u_i g_i v_i \neq 0.$$

Since $G(A)$ is a domain, it follows from Lemma 1.1, the definition of $\succeq_{gr}$ on $\sigma(\mathcal{B})$ (§1 (*)), Lemma 1.4 and CH.I Lemma 3.11 that

$$\sigma(f) = \sum_{d(\mathbf{LM}(u_i g_i v_i))=p} \lambda_i \sigma(u_i g_i v_i) = \sum_{d(\mathbf{LM}(u_i g_i v_i))=p} \lambda_i \sigma(u_i)\sigma(g_i)\sigma(v_i)$$

and

$$\begin{aligned}
\mathbf{LM}(\sigma(f)) = \sigma(\mathbf{LM}(f)) \;&\succeq_{gr}\; \sigma(\mathbf{LM}(u_i g_i v_i)) \\
&=\; \mathbf{LM}(\sigma(u_i g_i v_i)) \\
&=\; \mathbf{LM}(\sigma(u_i)\sigma(g_i)\sigma(v_i)).
\end{aligned}$$

This shows that $F = \sigma(f)$ has a Gröbner presentation by $\sigma(\mathcal{G})$, as desired.

(ii) Suppose that $G_i \in G(I)_{h_i} \subset G(A)_{h_i}$ (note that FI is the induced filtration), $i \in J$. For each G_i, chose $g_i \in F_{h_i}A \cap I - F_{h_i-1}A$ such that $G_i = \sigma(g_i)$. If $f \in F_pI - F_{p-1}I$, then $\sigma(f)$ has a Gröbner presentation, say $\sigma(f) = \sum_{i=1}^{s} \lambda_i \sigma(u_i)\sigma(g_i)\sigma(v_i)$, where $\lambda_i \in k-\{0\}$, $\sigma(u_i), \sigma(v_i) \in \sigma(\mathcal{B})$, with the property that $\mathbf{LM}(\sigma(f)) \succeq_{gr} \mathbf{LM}(\sigma(u_i)\sigma(g_i)\sigma(v_i))$ whenever $\sigma(u_i)\sigma(g_i)\sigma(v_i) \neq 0$. Since $G(A)$ is a domain, it follows from Lemma 1.4, the definition of $\succeq_{gr}$ on $\sigma(\mathcal{B})$ (§1 (*)) and CH.I Lemma 3.11 that

$$\begin{aligned}
\sigma(\mathbf{LM}(f)) = \mathbf{LM}(\sigma(f)) \;&\succeq_{gr}\; \mathbf{LM}(\sigma(u_i)\sigma(g_i)\sigma(v_i)) \\
&=\; \mathbf{LM}(\sigma(u_i g_i v_i)) \\
&=\; \sigma(\mathbf{LM}(u_i g_i v_i))
\end{aligned}$$

$$\mathbf{LM}(f) \succeq_{gr} \mathbf{LM}(u_i g_i v_i).$$

Note that $f - \sum_{i=1}^{s} \lambda_i u_i g_i v_i = h \in F_{p-1}A \cap I = F_{p-1}I$. After repeating the above procedure for a finite number of steps, a Gröbner presentation of f is obtained.

In the case where a left ideal L is considered, we only need to observe that $\sigma(f)$ may have a Gröbner presentation of the form $\sigma(f) = \sum_{i=1}^{s} \sigma(w_i)\sigma(g_i)$, where $\sigma(w_i) \in \sigma(\mathcal{B})$, and then use a similar argumentation as in the case of considering a two-sided ideal.   □

2.2. Theorem With notation as above, suppose that $(A, \mathcal{B}, \succeq_{gr})$ is a (left) admissible system where $\succeq_{gr}$ is some graded monomial ordering on $\mathcal{B}$, and that

$G(A)$ is a domain. Let I be a left ideal or two-sided ideal of A. The following statements hold.

(i) If $\mathcal{G} = \{g_i\}_{i \in J}$ is a (left) Gröbner basis for I, then $\widetilde{\mathcal{G}} = \{\widetilde{g}_i \mid g_i \in \mathcal{G}\}$ is a (left) Gröbner bsis for $\widetilde{I}$, with respect to the (left) admissible system $(\widetilde{A}, \widetilde{B}, \succeq_{gr})$ as determined in Theorem 1.5.

(ii) If $\{G_i\}_{i \in J}$ is a (left) Gröbner basis of $\widetilde{I}$ consisting of homogeneous elements, with respect to the (left) admissible system $(\widetilde{A}, \widetilde{B}, \succeq_{gr})$ as determined in Theorem 1.5, and if the g_i are choosen to be such that $X^{r_i}\widetilde{g}_i = G_i$ (indeed, each g_i is the image of G_i in A, see CH.I Lemma 3.11), $i \in J$, then $\mathcal{G} = \{g_i\}_{i \in J}$ is a (left) Gröbner basis for I.

Proof We prove the theorem for a two-sided ideal I (a similar argumentation works for a left ideal).

(i) We prove that every nonzero homogeneous element $F \in \widetilde{I}$ has a Gröbner presentation by $\widetilde{\mathcal{G}}$ in the sense of CH.II Definition 3.2. Suppose $F \in \widetilde{I}_p$. Then since FI is the induced filtration, $F = X^r \widetilde{f}$ for some $r \geq 0$ and some $f \in I$ (CH.I Lemma 3.11). Let $f = \sum_{i=1}^{s} \lambda_i u_i g_i v_i$ be a Gröbner presentation of f by $\mathcal{G}$, where $\lambda_i \in k - \{0\}$, $u_i v_i \in \mathcal{B}$, which has the property that $\mathbf{LM}(f) \succeq_{gr} \mathbf{LM}(u_i g_i v_i)$ whenever $u_i g_i v_i \neq 0$. By Lemma 1.1 and Lemma 1.4, $F = X^r \widetilde{f} = \sum \lambda_i X^r \widetilde{u_i g_i v_i} X^{r_i} = \sum \lambda_i X^r \widetilde{u}_i \widetilde{g}_i \widetilde{v}_i X^{r_i}$, where $0 \leq r_i \leq d(f)$. We prove that this is a Gröbner presentation of F by $\widetilde{\mathcal{G}}$, by considering two cases.

Case 1. $\mathbf{LM}(f) = \mathbf{LM}(u_i g_i v_i)$. Since $G(A)$ is a domain, it follows from Lemma 1.4 and CH.I Lemma 3.11 that

$$
\begin{aligned}
\mathbf{LM}(F) = \mathbf{LM}(X^r \widetilde{f}) &= X^r \mathbf{LM}(\widetilde{f}) = X^r (\mathbf{LM}(f))\widetilde{} \\
&= X^r (\mathbf{LM}(u_i g_i v_i))\widetilde{} = \mathbf{LM}(X^r \widetilde{u_i g_i v_i}) \\
&= \mathbf{LM}(X^r \widetilde{u}_i \widetilde{g}_i \widetilde{v}_i).
\end{aligned}
$$

Case 2. $\mathbf{LM}(f) \succ_{gr} \mathbf{LM}(u_i g_i v_i)$. By the definition of $\succeq_{gr}$ on $\widetilde{B}$ (§1 (*)) we have

$$
\widetilde{\mathbf{LM}(f)} X^h \succ_{gr} (\mathbf{LM}(u_i g_i v_i))\widetilde{} X^\ell, \quad h, \ell \in \mathbb{N}.
$$

Then, since $G(A)$ is a domain, it follows from the definition of $\succeq_{gr}$ on $\widetilde{B}$, Lemma 1.4 and CH.I Lemma 3.11 that

$$
\begin{aligned}
\mathbf{LM}(F) = \mathbf{LM}(X^r \widetilde{f}) = X^r \widetilde{\mathbf{LM}(f)} \quad &\succ_{gr} \quad X^r X^{r_i} (\mathbf{LM}(u_i g_i v_i))\widetilde{} \\
&= \mathbf{LM}(X^r X^{r_i} \widetilde{u_i g_i v_i}) \\
&= \mathbf{LM}(X^r \widetilde{u}_i \widetilde{g}_i \widetilde{v}_i X^{r_i}).
\end{aligned}
$$

This finishes the proof of (i).

(ii) Since the G_i are homogeneous elements in $\widetilde{I}$, we have $G_i = X^{r_i}\widetilde{g}_i$ for some $g_i \in I$ by CH.I Lemma 3.11. Note that since FI is the induced filtration, if $f \in I$ then $\widetilde{f}$ has a Gröbner presentation, say $\widetilde{f} = \sum_{i=1}^{s} \lambda_i X^{n_i} \widetilde{u}_i G_i \widetilde{v}_i X^{m_i}$, where $\lambda_i \in k - \{0\}$, $X^{n_i}\widetilde{u}_i, X^{m_i}\widetilde{v}_i \in \widetilde{B}$, with the property that $\mathbf{LM}(\widetilde{f}) \succeq_{gr}$

$\mathbf{LM}(X^{n_i}\widetilde{u}_i G_i \widetilde{v}_i X^{m_i}$ whenever $\widetilde{u}_i G_i \widetilde{v}_i \neq 0$. Substituting each G_i by $X^{r_i}\widetilde{g}_i$ and considering the image of $\widetilde{f}$ in A under the ring homomorphism $\widetilde{A} \to A$ by sending X to 1, we obtain $f = \sum_{i=1}^s \lambda_i u_i g_i v_i$ (see CH.I §3). We claim that this is a Gröbner presentation of f by $\mathcal{G} = \{g_i\}_{i \in J}$. Indeed, since $G(A)$ is a domain, it follows from Lemma 1.4 and CH.I Lemma 3.11 that

$$
\begin{aligned}
\widetilde{\mathbf{LM}(f)} = \mathbf{LM}(\widetilde{f}) \; &\succeq_{gr} \; \mathbf{LM}(X^{n_i}\widetilde{u}_i G_i \widetilde{v}_i X^{m_i}) \\
&= \; \mathbf{LM}(X^{n_i}\widetilde{u}_i X^{r_i}\widetilde{g}_i \widetilde{v}_i X^{m_i}) \\
&= \; X^{n_i+r_i+m_i}\widetilde{(\mathbf{LM}(u_i g_i v_i))} .
\end{aligned}
$$

By the definition of $\succeq_{gr}$ on $\widetilde{\mathcal{B}}$ we see that $\mathbf{LM}(f) \succeq_{gr} \mathbf{LM}(u_i g_i v_i)$, as claimed.$\square$

Remark Let A be a filtered ring with filtration FA and L a left ideal of A generated by $\{f_1, ..., f_s\}$. If we consider the filtration FL on L induced by FA and identify $G(L)$ with the graded left ideal $\oplus_n (F_n A \cap L + F_{n-1}A)/F_{n-1}A$ of $G(A)$, then generally $\{\sigma(f_1), ..., \sigma(f_s)\}$ does not generate $G(L)$ in $G(A)$. This had been a problem bothering people who are studying finitely generated algebras by the filtered-graded method. Now we see that this will be no longer a problem (even much better) when we are working on a (left) admissible system and using (left) Gröbner bases. For example, let us consider the left ideal L in the second Weyl algebra $A_2(k)$ generated by $f_1 = x_1 \partial_1$, $f_2 = x_2 \partial_1^2 - \partial_1$. Then one may check that $\{\partial_1\}$ is a Gröbner basis of L with respect to $x_1 >_{grlex} x_2 >_{grlex} \partial_1 >_{grlex} \partial_2$. However, with respect to the Bernstein filtration (standard filtration) on $A_2(k)$, it is obvious that $\sigma(f_1) = \sigma(x_1)\sigma(\partial_1)$ and $\sigma(f_2) = \sigma(x_2)\sigma(\partial_1)^2$ do not generate $G(L)$. For $\widetilde{A}$ and $\widetilde{L}$, a similar illustration is omitted.

3. Filtered-Graded Transfer of (Left) Dickson Systems

Let $A = k[a_i]_{i \in \Lambda}$, $\mathcal{B}$ a strictly filtered k-basis of A, and FA the $\mathcal{B}$-standard filtration on A. Let $(A, \mathcal{B}, \succeq_{gr})$ be a (left) admissible system, where $\succeq_{gr}$ is some graded monomial ordering on $\mathcal{B}$. Recall from CH.II §6 that the divisibility (CH.II Definition 2.2) yields another order $\succeq'_{gr}$ on $\mathcal{B}$ as follows:

If $w, u \in \mathcal{B}$ and $w \neq u$, then $u \succeq'_{gr} w$ if and only if $u = \mathbf{LM}(vws)$ for some $v, s \in \mathcal{B}$ with $v \neq 1$ or $s \neq 1$. (For a left admissible system, if and only if $u = \mathbf{LM}(vw)$ for some $v \in \mathcal{B}$ with $v \neq 1$);

and $(A, \mathcal{B}, \succeq_{gr})$ is called a (left) Dickson system if $\succeq'_{gr}$ is a Dickson partial ordering on $\mathcal{B}$ (CH.II Definition 6.4, Definition 6.7). In this section we discuss the filtererd-graded transfer of (left) Dickson systems associated with A, $G(A)$ and $\widetilde{A}$.

3.1. Theorem Let $(A, \mathcal{B}, \succeq_{gr})$ be a (left) admissible system where $\succeq_{gr}$ is some graded monomial ordering on $\mathcal{B}$. The following statements hold.
(i) Suppose that $G(A)$ is a domain. If $(A, \mathcal{B}, \succeq_{gr})$ is a (left) Dickson system, then so is the system $(G(A), \sigma(\mathcal{B}), \succeq_{gr})$ determined in Theorem 1.5.
(ii) If the system $(G(A), \sigma(\mathcal{B}), \succeq_{gr})$ determined in Theorem 1.5 is a (left) Dickson system, the so is $(A, \mathcal{B}, \succeq_{gr})$.

Proof (i) Let $N \subseteq \sigma(\mathcal{B})$ be a nonempty subset and put

$$N' = \left\{ w \in \mathcal{B} \mid \sigma(w) \in N \right\}.$$

By the definition of $\sigma(\mathcal{B})$ in §1 and Lemma 1.1, each element of N corresponds uniquely to an element in N'. Since $(A, \mathcal{B}, \succeq_{gr})$ is a (left) Dickson system, there exists a finite subset $W' \subseteq N'$ such that for each $u \in N'$ there is some $w \in W'$ with $u \succeq'_{gr} w$, i.e., $u = \mathbf{LM}(vws)$ for some $v, s \in \mathcal{B}$. Since $G(A)$ is a domain, it follows from Lemma 1.4 and CH.I Lemma 3.11 that

$$
\begin{aligned}
\sigma(u) &= \sigma(\mathbf{LM}(vws)) \\
&= \mathbf{LM}(\sigma(vws)) \\
&= \mathbf{LM}(\sigma(v)\sigma(w)\sigma(s)).
\end{aligned}
$$

Hence $\sigma(u) \succeq'_{gr} \sigma(w)$. Thus, for each element $\sigma(u) \in N$, there exists some $\sigma(w) \in W = \sigma(W') = \{\sigma(w) \mid w \in W'\} \subseteq N$ such that $\sigma(u) \succeq'_{gr} \sigma(w)$. Note that W is a finite subset of N, this proves that $(G(A), \sigma(\mathcal{B}), \succeq_{gr})$ is a (left) Dickson system.
(ii) Let $N \subseteq \mathcal{B}$ be a nonempty subset and put

$$N' = \left\{ \sigma(w) \mid w \in N \right\}.$$

Then N' is a nonempty subset of $\sigma(\mathcal{B})$. Since $(G(A), \sigma(\mathcal{B}), \succeq_{gr})$ is a (left) Dickson system, there exsists a finite subset $W' \subseteq N'$ such that for each $\sigma(u) \in N'$ there is some $\sigma(w) \in W'$ with $\sigma(u) \succeq'_{gr} \sigma(w)$, i.e., $\sigma(u) = \mathbf{LM}(\sigma(v)\sigma(w)\sigma(s))$ for some $\sigma(v), \sigma(s) \in \sigma(\mathcal{B})$. Note that $\sigma(u) \neq 0$ if and only if $\sigma(v)\sigma(w)\sigma(s) \neq 0$. It follows from Lemma 1.4 that

$$
\begin{aligned}
\sigma(u) &= \mathbf{LM}(\sigma(v)\sigma(w)\sigma(s)) \\
&= \mathbf{LM}(\sigma(vws)) \\
&= \sigma(\mathbf{LM}(vws)).
\end{aligned}
$$

Hence $u = \mathbf{LM}(vws)$ by Lemma 1.1, and consequently $u \succeq'_{gr} w$. Thus, for each element $u \in N$, there exists some $w \in W = \{w \in N \mid \sigma(w) \in W'\} \subseteq N$ such that $u \succeq'_{gr} w$. By Lemma 1.1, W is a finite subset of N. This proves that $(A, \mathcal{B}, \succeq_{gr})$ is a (left) Dickson system. $\square$

3.2. Theorem Let $(A, \mathcal{B}, \succeq_{gr})$ be a (left) admissible system where $\succeq_{gr}$ is some graded monomial ordering on $\mathcal{B}$. If the system $(\widetilde{A}, \widetilde{\mathcal{B}}, \succeq_{gr})$ determined in Theorem 1.5 is a (left) Dickson system, then so is $(A, \mathcal{B}, \succeq_{gr})$.

Proof Let $N \subseteq \mathcal{B}$ be a nonempty subset and put

$$N' = \left\{ \widetilde{w} \;\middle|\; w \in N \right\}.$$

Then N' is a nonempty subset of $\widetilde{\mathcal{B}}$. Since $(\widetilde{A}, \widetilde{\mathcal{B}}, \succeq_{gr})$ is a (left) Dickson system, there exsists a finite subset $W' \subseteq N'$ such that for each $\widetilde{u} \in N'$ there is some $\widetilde{w} \in W'$ with $\widetilde{u} \succeq'_{gr} \widetilde{w}$, i.e., $\widetilde{u} = \mathbf{LM}(X^n \widetilde{v} \widetilde{w} \widetilde{s} X^m)$ for some $X^n \widetilde{v}, X^m \widetilde{s} \in \widetilde{\mathcal{B}}$. Note that $\widetilde{u} \neq 0$ if and only if $\widetilde{v} \widetilde{w} \widetilde{s} \neq 0$. It follows from Lemma 1.4 and CH.I Lemma 3.11 that

$$
\begin{aligned}
\widetilde{u} &= \mathbf{LM}(X^n \widetilde{v} \widetilde{w} \widetilde{s} X^m) \\
&= X^{n+m+h} \mathbf{LM}(\widetilde{vws}) \\
&= X^{n+m+h} (\mathbf{LM}(vws))\widetilde{}.
\end{aligned}
$$

Hence $u = \mathbf{LM}(vws)$ by CH.I Proposition 3.8, and consequently $u \succeq'_{gr} w$. Thus, for each element $u \in N$, there exists some $w \in W = \{w \in N \mid \widetilde{w} \in W'\} \subseteq N$ such that $u \succeq'_{gr} w$. By CH.I Proposition 3.8, W is a finite subset of N. This proves that $(A, \mathcal{B}, \succeq_{gr})$ is a (left) Dickson system. $\square$

Remark (i) Let $A = k[a_i]_{i \in \Lambda}$ be an arbitrary k-algebra with the standard filtration FA. If A is Noetherian, then generally we hardly know whether $G(A)$ is Noetherian or not (on the contrary of CH.I Proposition 3.13). Nevertheless, Theorem 3.1(i) may be viewed as an algorithmic transfer of the (G-)Noetherian property of A to $G(A)$. (Note that the G-Noetherian property is indeed more stronger than the classical Noetherian Property, see CH.II §6).

(ii) It seems that an analogue of Theorem 3.1(i) does not exist for $\widetilde{A}$. The problem is that not every homogeneous element of $\widetilde{A}$ is of the form $\widetilde{f}$ with $f \in A$ (CH.I Lemma 3.11), and consequently the division transfer from A to $\widetilde{A}$ does not work well. Nevertheless, in the next section we will see that an analogue of Theorem 3.1(i) does exist for a large class of popular algebras (Theorem 4.2).

4. Filtered-Graded Transfer Applied to Quadric Solvable Polynomial Algebras

We finish this chapter by a closer look at the filtered-graded transfer applied to quadric solvable polynomial algebras.

Let $A = k[a_1, ..., a_n]$ be a quadric solvable polynomial algebra with the associated (left) admissible system $(A, \mathcal{B}, \succeq_{gr})$, as defined in CH.II §7 and CH.III §2, where

$$\mathcal{B} = \left\{ a^\alpha = a_1^{\alpha_1} \cdots a_n^{\alpha_n} \mid \alpha = (\alpha_1, ..., \alpha_n) \in I\!N^n \right\},$$

the set of all standard monomials in A, is a k-basis for A (and is obviously strictly filtrered), and $\succeq_{gr}$ is a graded monomial ordering on $\mathcal{B}$. Moreover, the generators of A satisfy the relations

$$a_j a_i = \lambda_{ji} a_i a_j + \sum_{l_0 \leq \ell} \lambda_{ji}^{k\ell} a_k a_\ell + \sum \lambda_h a_h + c_{ji}, \ 1 \leq i < j \leq n,$$

where $\lambda_{ji} \neq 0, \lambda_{ji}^{k\ell}, \lambda_h, c_{ji} \in k$.

4.1.   Theorem Let $(A, \mathcal{B}, \succeq_{gr})$ be as above, and let FA be the $\mathcal{B}$-standard filtration on A. The following holds.
(i) $G(A)$ is a domain.
(ii) $G(A)$ is a quadratic solvable polynomial algebra with the associated Dickson system $(G(A), \sigma(\mathcal{B}), \succeq_{gr})$. Moreover, if A is tame then so is $G(A)$.

Proof (i) Let F, G be homogeneous elements of degree p, q in $G(A)$. Then there are $f \in F_p A - F_{p-1} A$, $g \in F_q A - F_{q-1} A$ such that $\sigma(f) = F \neq 0, \sigma(g) = G \neq 0$. Supose that $\mathbf{LM}(f) = a^\alpha$, $\mathbf{LM}(g) = a^\beta$. Since $\mathcal{B}$ is strictly filtered, by Lemma 1.1 we have $d(\mathbf{LM}(f)) = |\alpha| = p$, $d(\mathbf{LM}(g)) = |\beta| = q$. By CH.II Proposition 7.2, we obtain $\mathbf{LM}(fg) = \mathbf{LM}(\mathbf{LM}(f)\mathbf{LM}(g)) = a^{\alpha+\beta}$ and and $d(a^{\alpha+\beta}) = |\alpha| + |\beta| = p + q$. Again, since $\mathcal{B}$ is strictly filtered, we have $a^{\alpha+\beta} \notin F_{p+q-1}A$ and hencce $fg \notin F_{p+q-1}A$. This shows that $\sigma(f)\sigma(g) = FG \neq 0$. Therefore, $G(A)$ is a domain.
(ii) This follows from the construction of $G(A)$, Theorem 1.5, part (i) of the present theorem , and Theorem 3.1(i).   □

4.2.   Theorem (a complement of Theorem 3.2) Let $(A, \mathcal{B}, \succeq_{gr})$ be as in Theorem 4.1. Then $\widetilde{A}$ is a quadratic solvable polynomial algebra with the associated Dickson system $(\widetilde{A}, \widetilde{\mathcal{B}}, \succeq_{gr})$.

Proof We prove that $\widetilde{A}$ satisfies the conditions of CH.II Proposition 7.2. By CH.I Proposition 3.16, $\widetilde{A} = k[\widetilde{a}_1, ..., \widetilde{a}_n, X]$, where X is the canonical homogeneous element of degree 1 represented by 1 in $\widetilde{A}_1 = F_1 A$. By Lemma 1.2 and Proposition 1.3, $\widetilde{\mathcal{B}} = \{\widetilde{a}_1^{\alpha_1} \cdots \widetilde{a}_n^{\alpha_n} X^\beta \mid (\alpha_1, ..., \alpha_n, \beta) \in I\!N^{n+1}\}$ forms a k-basis consisting of all standard monomials in $\widetilde{A}$. Since $G(A)$ is a domain, it follows from CH.I Lemma 3.11 that

$$
\begin{aligned}
\widetilde{a}_j \widetilde{a}_i = \widetilde{\overline{a_j}\,\overline{a_i}} &= \left(\lambda_{ji} a_i a_j + \sum_{k \leq \ell} \lambda_{ji}^{k\ell} a_k a_\ell + \sum \lambda_h a_h + c_{ji} \right)^{\widetilde{}} \\
&= \lambda_{ji} \widetilde{a}_i \widetilde{a}_j + \sum_{k \leq \ell} \lambda_{ji}^{k\ell} \widetilde{a}_k \widetilde{a}_\ell + \sum \lambda_h \widetilde{a}_h X + c_{ji} X^2 .
\end{aligned}
$$

Note that $X \widetilde{a}_j = \widetilde{a}_j X$, $j = 1, ..., n$, and that $\widetilde{a}_i \widetilde{a}_j \succ_{gr} \mathbf{LM}(F_{ij})$ by the definition of $\succeq_{gr}$ on $\widetilde{\mathcal{B}}$, where $F_{ij} = \sum_{k \leq \ell} \lambda_{ji}^{k\ell} \widetilde{a}_k \widetilde{a}_\ell + \sum \lambda_h \widetilde{a}_h X + c_{ji} X^2$. So we are done. $\square$

Furthermore, applying CH.III Proposition 1.5, Theorem 3.6, and Theorem 3.7 to a quadric solvable polynomial algebra, we fulfil the promise made after Corollary 3.8 in CH.III §3.

Let $A = k[a_1, ..., a_n]$ be a quadric solvable polynomial algebra with the associated (left) admissible system $(A, \mathcal{B}, \succeq_{gr})$. By the discussion of CH.III §2, A has the defining relations

$$
R_{ji} = X_j X_i - \lambda_{ji} X_i X_j - \sum_{k \leq \ell} \lambda_{ji}^{k\ell} X_k X_\ell - \sum \lambda_h X_h - c_{ji}, \ 1 \leq i < j \leq n.
$$

4.3. Proposition With notation as above, if, for $1 \leq i < j \leq n$, $\mathbf{LM}(R_{ji}) = X_j X_i$ with respect to $X_n >_{grlex} \cdots >_{grlex} X_1$. Then, with respect to the standard filtration FA on A, the following holds.

(i) The associated graded algebra $G(A)$ of A has the quadratic defining relations

$$
\sigma(R_{ji}) = X_j X_i - \lambda_{ji} X_i X_j - \sum \lambda_{ji}^{k\ell} X_k X_\ell, \ 1 \leq i < j \leq n,
$$

which form a Gröbner basis in the free algebra $k\langle X_1, ..., X_n \rangle$ with respect to $X_n >_{grlex} \cdots >_{grlex} X_1$.

(ii) The Rees algebra $\widetilde{A}$ of A has the quadratic defining relations

$$
\begin{aligned}
& T X_i - X_i T, \ 1 \leq i \leq n, \\
& \widetilde{R}_{ji} = X_j X_i - \lambda_{ji} X_i X_j - \sum \lambda_{ji}^{k\ell} X_k X_\ell - \sum \lambda_h X_h T - c_{ji} T^2, \ 1 \leq i < j \leq n,
\end{aligned}
$$

which form a Gröbner basis in the free algebra $k\langle X_1, ..., X_n, T \rangle$ with respect to $X_n >_{grlex} \cdots >_{grlex} X_1 >_{grlex} T$.

$$\square$$

At this stage, we mention the following application of the foregoing results to those examples which have been considered before (CH.I §5, CH.II §7, CH.III §3).

4.4. Corollary Let A be one of the following k-algebras:

(i) The nth Weyl algebra $A_n(k)$;

(ii) The additive analogue $A_n(q_1, ..., q_n)$ of $A_n(k)$;

(iii) The multiplicative analogue $\mathcal{O}_n(\lambda_{ji})$ of $A_n(k)$;

(iv) The enveloping algebra $U(\mathbf{g})$ of an n-dimensional k-Lie algebra $\mathbf{g}$;

(v) The nth q-heisenberg algebra $\mathbf{h}_n(q)$;

(vi) Any quadric solvable polynomial algebra constructed in CH.III §2.

Let $(A, \mathcal{B}, \geq_{grlex})$ be the (left) Dickson system associated with A, where $\mathcal{B}$ is the standard k-basis of A consisting of all standard monomials and $\geq_{grlex}$ is the graded lexicographic ordering that is suitably defined on $\mathcal{B}$. Consider the $\mathcal{B}$-standard filtration on A. Then $(G(A), \sigma(\mathcal{B}), \geq_{grlex})$ and $(\widetilde{A}, \widetilde{\mathcal{B}}, \geq_{lexgr})$ are (left) Dickson systems, and hence A, $G(A)$ and $\widetilde{A}$ are all (left) G-Noetherian.

CHAPTER V
GK-dimension of Modules over Quadric Solvable Polynomial Algebras and Elimination of Variables

A well-known theme in commutative computational algebra is to find the symbolic solutions of polynomial equations, or in other words, is the elimination of variables, and on this aspect, the Wu's method and Gröbner basis method have been quite powerful (cf. [BW], [Cho], [CLO'], [Wan]). In the elimination theory using Gröbner bases, a notable fact is that there is a nice relation between the number of eliminated variables and the dimension of the algebraic variety determined by the system of polynomial equations. More precisely, let $A = k[x_1, ..., x_n]$ be the commutative polynomial k-algebra in n variables and I a proper ideal of A such that the dimension of the affine algebraic variety $V(I)$ is d. Then it follows from ([LVO5] CH.V §7) that

- for every subset $U = \{x_{i_1}, ..., x_{i_{d+1}}\} \subset \{x_1, ..., x_n\}$, $k[x_{i_1}, ..., x_{i_{d+1}}] \cap I \neq \{0\}$, i.e., there is a nonzero member of I that only depends on the generators in U. In particular,

$$k[x_1, ..., x_d, x_{d+i}] \cap I \neq \{0\}, \ i = 1, ..., n - d.$$

At this stage, the wonderful thing is that, once a Gröbner basis is produced, the computation of the dimension of $V(I)$ and the elimination process can be algorithmically realized simultaneously (e.g., see [BW], [CLO']).

Since the natural noncommutative algebraic translation of the dimension of an algebraic variety is the Gelfand-Kirillov dimension (GK-dimension) of a module over a k-algebra (e.g., see [KL]), it is natural to ask

Question Can we use Gröbner bases to compute GK-dimension for modules over a certain kind of algebra and have an analogue of the above (•)?

Let $A = k[a_1, ..., a_n]$ be a *quadric* solvable polynomial algebra as defined in in CH.II §7 and CH.III §2, and let L be a left ideal of A with the left Gröbner basis $\mathcal{G} = \{g_1, ..., g_s\}$. For each g_j with the leading monomial

$$\mathbf{LM}(g_j) = a_1^{\alpha_{j1}} a_2^{\alpha_{j2}} \cdots a_n^{\alpha_{jn}}, \quad j = 1, ..., s,$$

we associate g_j to the n-tuple of exponents

$$R_j = (\alpha_{j1}, ..., \alpha_{jn}) \in I\!\!N^n, \quad j = 1, ..., s.$$

In this chapter, using the filtered-graded transfer of Gröbner bases developed in CH.IV, we first demonstrate how to compute the GK-dimension of the cyclic left module $M = A/L$ (and hence of any finitely generated left module) in the case where A is *linear* (Ch.III Definition 2.1), by manipulating only the data $(R_1, R_2, ..., R_s)$ actually as in the *commutative* case; and furthermore, we use a double filtered-graded transfer of Gröbner bases linked by the $\succeq_{gr}$-filtration (see §6 for the definition) to obtain the same result for general quadric solvable polynomial algebras. This, in turn, yields an elimination (of variables) lemma for quadric solvable polynomial algebras. In other words, we have a complete answer to the above question for the class of quadric solvable polynomial algebras.

1. Gröbner Bases in Homogeneous Solvable Polynomial Algebras

We start with homogeneous solvable polynomial algebras.

Let $A = k[a_1, ..., a_n]$ be a homogeneous solvable polynomial algebra in the sense of Ch.III Definition 2.1, and let $(A, \mathcal{B}, \succeq_{gr})$ be the associated (left) admissible system of A, where

$$\mathcal{B} = \left\{ a^\alpha = a_1^{\alpha_1} \cdots a_n^{\alpha_n} \;\middle|\; \alpha = (\alpha_1, ..., \alpha_n) \in I\!\!N^n \right\}$$

is the standard k-basis of A and $\succeq_{gr}$ is a graded monomial ordering on $\mathcal{B}$. Then A is a Noetherian domain (CH.II §7).

In this section, we record briefly some easily verified properties of Gröbner bases for left ideals in A (so Gröbner basis is refered to *left* Gröbner basis).   Since our discussion will not only deal with the monomials in $\mathcal{B}$ but will also deal with the coefficient of the product of two monomials in computation, due to the noncommutativity, we write $\mathbf{T}(A)$ for the set of all *terms* in A, i.e.,

$$\mathbf{T}(A) = \left\{ ca^\alpha \mid c \in k, \, a^\alpha \in \mathcal{B} \right\}.$$

Moreover, for a nonzero element $f \in A$ with

$$f = \sum c_\alpha a^\alpha, \, c_\alpha \in k - \{0\}, \, a^\alpha \in \mathcal{B},$$

we write

$$\mathbf{LT}(f) = \mathbf{LC}(f)\mathbf{LM}(f)$$

for the *leading term* of f (see CH.II Definition 1.1).

1.1. Lemma (i) A is a positively graded k-algebra, i.e., $A = \oplus_{m \in \mathbb{N}} A_m$ with $A_m = \{\sum_{|\alpha|=m} c_\alpha a^\alpha \mid a^\alpha = a_1^{\alpha_1} \cdots a_n^{\alpha_n}, |\alpha| = \alpha_1 + \cdots + \alpha_n\}$, in particular, $A_0 = k$, $A_1 = \sum_{i=1}^n ka_i$.
(ii) If $c_\alpha a^\alpha, c_\beta a^\beta \in \mathbf{T}(A)$, then $c_\beta a^\beta$ is divisible by $c_\alpha a^\alpha$ (from left hand side, see CH.II Definition 2.2) if and only if there is some $c_\gamma a^\gamma \in \mathbf{T}(A)$ such that $c_\beta a^\beta = c_\alpha c_\gamma a^\gamma a^\alpha$.

$\square$

Next, we deal with monomial left ideals. A left ideal L of A is said to be a *monomial left ideal* if it has a generating set consisting of standard monomials in $\mathcal{B}$, i.e., $L = \sum_{\alpha \in \Lambda} Aa^\alpha$, $\Lambda \subset \mathbb{N}^n$, $a^\alpha \in \mathcal{B}$. By Lemma 1.1, the following two properties of a monomial left ideal are derived as for a commutative monomial ideal in the commutative polynomial algebra $k[x_1, ..., x_n]$.

1.2. Lemma Let $L = \sum_{\alpha \in \Lambda} Aa^\alpha$ be a monomial left ideal of A. Then a standard monomial $a^\beta \in L$ if and only if a^β is divisible by a^α for some $\alpha \in \Lambda$.

$\square$

1.3. Lemma Let L be a monomial left ideal of A, and let $f \in A$. Then the following are equivalent:
(i) $f \in L$;
(ii) Every term of f lies in L;
(iii) f is a k-linear combination of the standard monomials in $L \cap \mathcal{B}$.

$\square$

Let L be a left ideal of A, and $\mathbf{LT}(L)$ the set of leading terms of elements of L. We denote by $\langle \mathbf{LT}(L)]$ the monomial left ideal generated by $\mathbf{LT}(L)$ in A. If furthermore $\mathcal{G} = \{g_1, ..., g_s\}$ is a left Gröbner basis of L, then we denote by $\langle \mathbf{LT}(\mathcal{G})]$ the monomial left ideal generated by $\mathbf{LT}(\mathcal{G}) = \{\mathbf{LT}(g_1), ..., \mathbf{LT}(g_s)\}$ in

A. Generally, in a solvable polynomial algebra we do not have $\langle \mathbf{LT}(L) \rangle =$ $\langle \mathbf{LT}(\mathcal{G}) \rangle]$ because of CH.II Proposition 4.2. However, for a homogeneous solvable polynomial algebra A, it follows from Lemma 1.2 and Lemma 1.3 that the following result holds as in the commutative case.

1.4. Proposition Let L be a left ideal of A and $\mathcal{G} = \{g_1, ..., g_s\} \subset L$.
(i) (compare with CH.II Proposition 4.1) If $L = \langle a^\alpha \in \mathcal{B} \mid \alpha \in \Lambda \subset I\!N^n \rangle]$ is a monomial left ideal, then $L = \langle \mathbf{LT}(L) \rangle]$ and $\{a^\alpha \mid \alpha \in \Lambda\}$ is a Gröbner basis for L.
(ii) (compare with CH.II Proposition 4.2) $\mathcal{G} = \{g_1, ..., g_s\}$ is a Gröbner basis of L if and only if $\langle \mathbf{LT}(L) \rangle] = \langle \mathbf{LT}(\mathcal{G}) \rangle]$.
(iii) (compare with CH.II Theorem 3.1) Let $\mathcal{G}$ be a Gröbner basis of L, then A/L has a k-basis $\{[a^\alpha] \mid a^\alpha \in \mathcal{B} - \langle \mathbf{LT}(\mathcal{G}) \rangle]\}$, where $[a^\alpha]$ is the class of a^α in A/L.

□

2. The Hilbert Function of A/L

Let $A = k[a_1, ..., a_n]$ be a homogeneous solvable polynomial algebra with the associated (left) admissible system $(A, \mathcal{B}, \succeq_{gr})$, and let L be a left ideal of A. The aim of this section is to introduce the Hilbert function HF_L of the A-module A/L, and reduce the study of HF_L to the study of $HF_{\langle \mathbf{LT}(L) \rangle]}$.

Consider the standard filtration FA on A:

$$F_m A = \left\{ \sum c_\alpha a^\alpha \in A \;\middle|\; a^\alpha \in \mathcal{B}, \; |\alpha| \leq m \right\}, \quad m \in I\!N,$$

in which $F_0 A = k$, and each $F_m A$ is a finite dimensional k-space. Then FA induces the filtration FL on L: $F_m L = L \cap F_m A$, $m \in I\!N$. The *Hilbert function* of the A-module A/L, denoted HF_L, is defined by putting

$$HF_L(m) = \dim_k \left(\frac{F_m A}{F_m L} \right), \quad m \in I\!N.$$

The *key* idea making the Hilbert function computable comes from the following noncommutative version of a celebrated result obtained from (F.S. Macaulay, Some properties of enumeration in the theory of modular systems, *Proc. London math. Soc.*, 26(1927), 531–555). The next proposition will be recaptured in a more general setting in later §7, and the proof given below also shows why *a graded monomial ordering is necessary*.

2.1. Proposition Let $(A, \mathcal{B}, \succeq_{gr})$, L be as given above, and let $\langle \mathbf{LT}(L)]$ be the left ideal generated by $\mathbf{LT}(L)$, the set of leading terms of all elements in L. Then

$$HF_L(m) = HF_{\langle \mathbf{LT}(L)]}(m), \quad m \in I\!N.$$

Proof Taking a left Gröbner basis $\mathcal{G}$ of L and putting $\mathcal{B}_{\leq m} = \mathcal{B} \cap F_m A$, we claim that there are k-space isomorphisms:

$$\frac{F_m A}{F_m L} \xrightarrow{\cong} k\text{-Span}\left\{a^\alpha \mid a^\alpha \in \mathcal{B}_{\leq m} - \langle \mathbf{LT}(L)]\right\} \xleftarrow{\cong} \frac{F_m A}{F_m \langle \mathbf{LT}(L)]}$$

Since $\langle \mathbf{LT}(L)]$ is a monomial left ideal, the second isomorphism is easily obtained by Lemma 1.3 and Proposition 1.4. We will obtain the first isomorphism by showing that the k-spaces $F_m L$ and $F_m \langle \mathbf{LT}(L)]$ have the same dimension. First note that

$$(1) \qquad \left\{ \mathbf{LM}(f) \;\middle|\; f \in F_m L \right\} = \{ \mathbf{LM}(f_1), ..., \mathbf{LM}(f_s) \}$$

for a finite number of $f_1, ..., f_s \in F_m L$. By rearranging and deleting duplicates, we may assume that

$$(2) \qquad \mathbf{LM}(f_1) \succeq_{gr} \mathbf{LM}(f_2) \succeq_{gr} \cdots \succeq_{gr} \mathbf{LM}(f_s).$$

We claim that $\{f_1, ..., f_s\}$ forms a basis of $F_m L$ as a k-space. To see this, consider a nontrivial linear combination $c_1 f_1 + \cdots + c_s f_s$ and choose the smallest i such that $c_i \neq 0$. It follows from (2) above that no $c_i \mathbf{LT}(f_i)$ can be canceled, and hence the linear combination is nonzero. Thus, $f_1, ..., f_s$ are linearly independent. Next, let $W = k\text{-Span}\{f_1, ..., f_s\} \subset F_m L$. If $W \neq F_m L$, pick $f \in F_m L - W$ with $\mathbf{LM}(f)$ minimal (note that $\succeq_{gr}$ is a well-ordering). By (1), $\mathbf{LM}(f) = \mathbf{LM}(f_i)$ for some i, and hence, $\mathbf{LT}(f) = \lambda \mathbf{LT}(f_i)$ for some $\lambda \in k$. Then $f - \lambda f_i \in F_m L$ has a smaller leading monomial, so that $f - \lambda f_i \in W$ by the minimality of $\mathbf{LM}(f)$. This implies $f \in W$, which is a contradiction. It follows that $W = F_m L$, and we conclude that $\{f_1, ..., f_s\}$ forms a k-basis. A similar argumentation as above shows that $\mathbf{LM}(f_1), ..., \mathbf{LM}(f_s)$ are k-linearly independent. Note that we are using $\succeq_{gr}$. This means that for any nonzero $f \in A$, if $\mathbf{LM}(f) \in F_m A$, then so does f (CH.IV Lemma 1.4). It follows immediately from Lemma 1.3 and the above (1) that

$$F_m \langle \mathbf{LT}(L)] = k\text{-Span}\{\mathbf{LM}(f_1), ..., \mathbf{LM}(f_s)\}.$$

So we are done. $\qquad \square$

Now, recall from Lemma 1.1(i) that $A = k[a_1, ..., a_n]$ is positively graded where $\deg(a_i) = 1$, $i = 1, ..., n$, i.e., $A = \oplus_{m \in I\!N} A_m$ with $A_m = \{\sum c_\alpha a^\alpha \in A \mid a^\alpha \in \mathcal{B}, |\alpha| = m\}$, and each A_m is a finite dimensional k-space. Let L be a *graded left ideal* of A. Then $L = \oplus_{m \in \mathbb{N}} L_m$ with $L_m = A_m \cap L$. The *Hilbert function* of the graded A-module A/L, denoted HF_L^g, is defined by putting

$$HF_L^g(m) = \dim_k \left(\frac{A_m}{L_m} \right), \quad m \in \mathbb{N}.$$

2.2. Proposition With notation as above,

$$HF_L^g(m) = HF_L(m) - HF_L(m-1), \quad m \in \mathbb{N}.$$

Proof Since in this case we have $F_m A = \oplus_{p \leq m} A_p$, $m \in \mathbb{N}$, there is a natural isomorphism of k-spaces:

$$\frac{A_m}{L_m} \xrightarrow{\cong} \frac{F_m A + L}{F_{m-1} A + L}$$

It follows that $\dim_k(A_m/L_m) = \dim_k(F_m A/F_m L) - \dim_k(F_{m-1}A/F_{m-1}L)$, as desired. $\square$

3. The Hilbert Polynomial of A/L

Let $(A, \mathcal{B}, \succeq_{gr})$ be the (left) admissible system associated with a homogeneous solvable polynomial algebra A, and let L be a left ideal of A with the Hilbert function HF_L as defined in §2. Based on Proposition 2.1 of last section, in this section we prove that there exists a polynomial $h_L(x) \in \mathbb{Q}[x]$ with positive leading coefficient such that, for $m \gg 0$, $HF_L(m) = h_L(m)$. We approach this by mimicking the commutative case [BW], where the main idea is to

- classify $S = \mathcal{B}_{\leq m} - \langle \mathbf{LT}(L) \rangle$, where $\mathcal{B}_{\leq m} = \mathcal{B} \cap F_m A$, by a suitably defined equivalence relation on S, and then
- compute the number of elements of each equivalence class.

To this end, we need a few more technicalities.

First note that $\mathbf{LM}(ts) = a^{\alpha+\beta} = \mathbf{LM}(st)$ for $t = a^\alpha$, $s = a^\beta \in \mathcal{B}$. Next, for $t = a_1^{\alpha_1} \cdots a_n^{\alpha_n} \in \mathcal{B}$, $M \in \mathbb{N}$, we write $\mathbf{top}_M(t)$ for the set consisting of all indices $i_j \in \{1, ..., n\}$ with $\alpha_{i_j} \geq M$. Suppose $\mathbf{top}_M(t) = \{i_1, ..., i_s\}$ with $i_1 < \cdots < i_s$. We define

$$\mathbf{sh}_M(t) = \left(a_{i_1}^M \cdots a_{i_s}^M \right) \left(a_{i_{s+1}}^{\alpha_{i_{s+1}}} \cdots a_{i_n}^{\alpha_{i_n}} \right),$$

where $i_{s+1} < \cdots < i_n$. Since A is homogeneous, we have $\mathbf{sh}_M(t) \in \mathbf{T}(A)$ (see §1). Then it is clear that

$$(*) \qquad\qquad \mathbf{sh}_M(t) = \mathbf{sh}_M(\mathbf{LM}(\mathbf{sh}_M(t))).$$

3.1. Lemma With notation as above, let $S \subset \mathcal{B}$ be a subset and $M \in I\!N$. The following hold.

(i) The relation $\sim_M$ defined by

$$s \sim_M t \quad \text{if and only if } \mathbf{sh}_M(s) = \mathbf{sh}_M(t)$$

is an equivalence relation on S.

(ii) Let $[s]_{\sim_M}$ denote the equivalence class of $s \in S$. If $\mathbf{LM}(\mathbf{sh}_M(s)) \in S$ for all $s \in S$, then the set

$$R_M = \left\{ t \in S \mid \mathbf{LM}(\mathbf{sh}_M(t)) = t \right\}$$

is a system of unique representatives for the partition of S into equivalence classes with respect to $\sim_M$. The set R_M can also be described as

$$R_M = \left\{ t \in S \mid t = a_1^{\alpha_1} \cdots a_n^{\alpha_n},\ \alpha_i \le M,\ 1 \le i \le n \right\}.$$

(iii) Let R_M be as in (ii). For $t \in R_M$, we have

$$[t]_{\sim_M} = \left\{ \mathbf{LM}(st) \mid \mathbf{LM}(st) \in S,\ s = a_1^{\gamma_1} \cdots a_n^{\gamma_n} \text{ with } \gamma_i = 0 \text{ for } i \notin \mathbf{top}_M(t) \right\}$$

Proof (i) This is clear.

(ii) By the assumption we have $\mathbf{LM}(\mathbf{sh}_M(s)) \in S$ for all $s \in S$. Hence $s \sim_M \mathbf{LM}(\mathbf{sh}_M(s))$ by the above $(*)$, i.e., $[s]_{\sim_M} = [\mathbf{LM}(\mathbf{sh}_M(s))]_{\sim_M}$ with $\mathbf{LM}(\mathbf{sh}_M(s)) \in R_M$. If $s \sim_M t$, then $\mathbf{sh}_M(s) = \mathbf{sh}_M(t)$. This means that $\mathbf{LM}(\mathbf{sh}_M(s)) = \mathbf{LM}(\mathbf{sh}_M(t))$, or in other words, the representative of $[s]_{\sim_M}$ in R_M is unique. Furthermore, if $t \in R_M$, $t = a_1^{\alpha_1} \cdots a_n^{\alpha_n}$ and $\mathbf{sh}_M(t) = (a_{i_1}^{M} \cdots a_{i_s}^{M})(a_{i_{s+1}}^{\alpha_{i_{s+1}}} \cdots a_{i_n}^{\alpha_{i_n}})$, then clearly $\alpha_i \le M$ for $1 \le i \le n$.

(iii) Let $t \in R_M$, $t = a_1^{\alpha_1} \cdots a_n^{\alpha_n}$. Suppose $t' = a_1^{\beta_1} \cdots a_n^{\beta_n} \in S$ and $t' \sim_M t$. Then $\mathbf{sh}_M(t') = \mathbf{sh}_M(t)$, i.e., $\mathbf{sh}_M(t) = (a_{i_1}^{M} \cdots a_{i_s}^{M})(a_{i_{s+1}}^{\alpha_{i_{s+1}}} \cdots a_{i_n}^{\alpha_{i_n}}) = \mathbf{sh}_M(t')$. Since $\mathbf{LM}(\mathbf{sh}_M(t)) = t$, for $i_k \notin \mathbf{top}_M(t)$ we have $\beta_{i_k} = \alpha_{i_k} < M$; for $i_k \in \mathbf{top}_M(t)$ we have $\beta_{i_k} \ge M = \alpha_{i_k}$ by (ii). Hence there exists $s = a_1^{\gamma_1} \cdots a_n^{\gamma_n}$ with $\gamma_i = 0$ for $i \notin \mathbf{top}_M(t)$, such that $t' = \mathbf{LM}(st)$.

Conversely, for any $s = a_1^{\gamma_1} \cdots a_n^{\gamma_n}$ with $\gamma_i = 0$, $i \notin \mathbf{top}_M(t)$, if $\mathbf{LM}(st) \in S$ then it is clear that $t \sim_M \mathbf{LM}(st)$. $\square$

3.2. Lemma Let L be a left ideal of A and $\mathcal{G}$ a Gröbner basis of L. If we put

$$M = \max \left\{ \beta_i \mid \mathbf{LT}(g) = c_\beta a_1^{\beta_1} \cdots a_n^{\beta_n},\ g \in \mathcal{G},\ 1 \le i \le n \right\},$$

then for every $t \in \mathcal{B}$, whenever $i \in \mathbf{top}_M(t)$ and $\nu \in I\!N$, $t \in \mathcal{B} - \langle \mathbf{LT}(L)]$ if and only if $\mathbf{LM}(a_i^\nu t) \in \mathcal{B} - \langle \mathbf{LT}(L)]$.

Proof Writing $t = a_1^{\alpha_1} \cdots a_n^{\alpha_n}$, $\mathbf{top}_M(t) = \{i_1, ..., i_s\}$, then

$$\mathbf{sh}_M(t) = \left(a_{i_1}^{M} \cdots a_{i_s}^{M} \right) \left(a_{i_{s+1}}^{\alpha_{i_{s+1}}} \cdots a_{i_n}^{\alpha_{i_n}} \right).$$

If $\mathbf{LM}(a_i^\nu t) \in \mathcal{B} - \langle \mathbf{LT}(L)\rangle$, then $t \in \mathcal{B} - \langle \mathbf{LT}(L)\rangle$. Conversely, let $t \in \mathcal{B} - \langle \mathbf{LT}(L)\rangle$. Assume the contrary that $\mathbf{LM}(a_i^\nu t) \in \langle \mathbf{LT}(L)\rangle$. Then $a_i^\nu t$ is divisible by some $\mathbf{LT}(g) = c_\beta a_1^{\beta_1} \cdots a_n^{\beta_n}$. Since $i \in \mathbf{top}_M(t)$, we have $\alpha_i \geq M \geq \beta_i$. It follows that t is divisible by $\mathbf{LT}(g)$, i.e., $t \in \langle \mathbf{LT}(L)\rangle$ (see Lemma 1.1(ii)), a contradiction. Hence $\mathbf{LM}(a_i^\nu t) \in \mathcal{B} - \langle \mathbf{LT}(L)\rangle$. $\qquad\square$

Before proving the main result, let us recall a fact from Proposition 2.1 and its proof:

$$(\triangle) \qquad \begin{aligned} HF_L(m) &= HF_{\langle \mathbf{LT}(L)\rangle}(m) \\ &= |\mathcal{B}_{\leq m} - \langle \mathbf{LT}(L)\rangle|, \quad m \in I\!N, \end{aligned}$$

where $\mathcal{B}_{\leq m} = \mathcal{B} \cap F_m A$.

3.3. Theorem Let $(A, \mathcal{B}, \succeq_{gr})$ and L be as before. Let $\mathcal{G}$ be a left Gröbner basis of L (note that A is solvable), and put

$$M = \max\left\{ \beta_i \;\middle|\; \mathbf{LT}(g) = c_\beta a_1^{\beta_1} \cdots a_n^{\beta_n}, \; g \in \mathcal{G}, \; 1 \leq i \leq n \right\}.$$

Then there exists a unique polynomial $h_L(x) \in \mathbb{Q}[x]$ with positive leading coefficient such that

$$HF_L(m) = h_L(m), \quad \text{for all } m \geq n \cdot M.$$

If the ground field is computable, then $h_L(x)$ and the number $n \cdot M$ can be computed from any given generating set of L.

Proof By the previous formula $(\triangle)$, the desired polynomial $h_L(x)$ must satisfy

$$(1) \qquad\qquad h_L(m) = |\mathcal{B}_{\leq m} - \langle \mathbf{LT}(L)\rangle|$$

for all $m \geq n \cdot M$. We will arrive at such a polynomial by counting the elements of $\mathcal{B}_{\leq m} - \langle \mathbf{LT}(L)\rangle$. To this end, we let $m \in I\!N$ with $m \geq n \cdot M$. It follows from Lemma 3.2 that $S = \mathcal{B}_{\leq m} - \langle \mathbf{LT}(L)\rangle$ satisfies $\mathbf{LM}(\mathbf{sh}_M(s)) \in S$ for all $s \in S$. Hence S is the disjoint union of the equivalence classes with respect to the equivalence relation $\sim_M$ of Lemma 3.1(i). Using the set R_M of Lemma 3.1(ii) as a system of unique representatives, we have

$$(2) \qquad\qquad |\mathcal{B}_{\leq m} - \langle \mathbf{LT}(L)\rangle| = \sum_{t \in R_M} |[t]_{\sim_M}|,$$

where $[t]_{\sim_M}$ is the equivalence class of $t \in R_M$.

Now from Lemma 3.1(iii) and Lemma 3.2 it is clear that for every $t \in R_M$, if $t = a_1^{\alpha_1} \cdots a_n^{\alpha_n}$, then

$$[t]_{\sim_M} = \left\{ \mathbf{LM}(st) \;\middle|\; s = a_1^{\gamma_1} \cdots a_n^{\gamma_n}, \; \gamma_i = 0, \text{ for } i \notin \mathbf{top}_M(t), |\gamma| \leq m - |\alpha| \right\}.$$

It follows that

$$|[t]_{\sim M}| = \begin{pmatrix} m - |\alpha| + |\mathbf{top}_M(t)| \\ |\mathbf{top}_M(t)| \end{pmatrix}$$

which is a polynomial of degree $|\mathbf{top}_M(t)|$ in $m - |\alpha|$. Combining this with (2), we have obtained a polynomial $h_L(x) \in \mathbb{Q}[x]$ with positive leading coefficient that satisfies (1) and has

$$\deg h_L(x) = \max\left\{|\mathbf{top}_M(t)| \,\middle|\, t \in R_M\right\}$$

$$= \max\left\{|\mathbf{top}_M(s)| \,\middle|\, s \in \mathcal{B}_{\leq m} - \langle \mathbf{LT}(L)\rangle\right\}.$$

Since by Proposition 1.4 we have $\langle \mathbf{LT}(L)\rangle = \langle \mathbf{LT}(\mathcal{G})\rangle$, the existence proof of the polynomial $h_L(x)$ that we have just given shows that the last statement of the theorem concerning computability is true. It is also clear that there can be only one $h_L(x) \in \mathbb{Q}[x]$ satisfying $HF_L(m) = h_L(m)$ for infinitely many $m \in \mathbb{N}$. $\square$

3.4. Definition The polynomial $h_L(x)$ obtained in the above theorem is called the *Hilbert polynomial* of A/L.

Finally, since $A = k[a_1, ..., a_n]$ is positively graded (see Lemma 1.1) with $\deg(a_i) = 1$, $i = 1, ..., n$, if L be a graded left ideal of A, then Proposition 2.2 enables us to mention the following result.

3.5. Theorem With notation as in §2, there exists a unique polynomial in $\mathbb{Q}[x]$ with positive leading coefficient, which is called the *Hilbert polynomial* of the graded A-module A/L and is denoted by $h_L^g(x)$, such that for $m \gg 0$

$$HF_L^g(m) = \dim_k\left(\frac{A_m}{L_m}\right) = h_L^g(m).$$

Moreover, if $h_L(x)$ has degree d then $h_L^g(x)$ has degree $d - 1$.

$\square$

4. GK-dimension Computation and Elimination of Variables (Homogeneous Case)

Let $A = k[a_1, ..., a_n]$, $(A, \mathcal{B}, \succeq_{gr})$ and L be as in §3. The argumentation of previous sections has indeed shown that the degree of the Hilbert polynomial $h_L(x)$ of the A-module A/L is nothing but the Gelfand-Kirillov dimension of

A/L (e.g., see [KL]) which is usually denoted by GK.dim(A/L), i.e.,

$$\deg h_L(x) = \text{GK.dim}(A/L).$$

In this section, we show that there is an algorithm for computing GK.dim(A/L) directly without computing $h_L(x)$; and we also show that a noncommutative analogue of the result ($\bullet$) (as mentioned in the beginning of this chapter) exists for homogeneous solvable polynomial algebras.

As in the commutative case (e.g., [BW], [CLO']), we first show that $\deg h_L(x)$ is closely related to the "independence" of generators of A (modulo L).

If $U = \{a_{i_1}, ..., a_{i_r}\} \subset \{a_1, ..., a_n\}$ with $i_1 > \cdots > i_r$, then the subalgebra $k[U] = k[a_{i_1}, ..., a_{i_r}]$ of A generated by U and k is also a homogeneous solvable polynomial k-algebra. Let $\mathbf{T}(U)$ be the set of of all terms in $k[U]$. We say that U is *independent* (modulo L), if $\mathbf{T}(U) \cap \langle \mathbf{LT}(L) \rangle = \{0\}$.

4.1. Proposition With notation as above, put

$$d = \max\Big\{ |U| \ \Big| \ U \subset \{a_1, ..., a_n\} \text{ independent (modulo } L) \Big\}.$$

Then $\deg h_L(x) = d$.

Proof Let $\mathcal{G}$ be a Gröbner basis of L. First recall from §3 that for

$$M = \max\Big\{ \beta_i \ \Big| \ \mathbf{LT}(g) = c_\beta a_1^{\beta_1} \cdots a_n^{\beta_n}, \ g \in \mathcal{G}, \ 1 \leq i \leq n \Big\},$$

we have

$$\deg h_L(x) = \max\Big\{ |\mathbf{top}_M(t)| \ \Big| \ t \in \mathcal{B}_{\leq m} - \langle \mathbf{LT}(L) \rangle, \ m \geq n \cdot M \Big\}.$$

To prove the inequality "$\leq$", assume for a contradiction that there exists $t \in \mathcal{B}_{\leq m} - \langle \mathbf{LT}(L) \rangle$, $t = a_1^{\alpha_1} \cdots a_n^{\alpha_n}$ with $\alpha_i \geq M$ for more than d many indices. Then there exists a subset $U = \{a_{i_1}, ..., a_{i_r}\} \subset \{a_1, ..., a_n\}$ with $i_1 > \cdots > i_r$, $r > d$, and a decomposition $t = t_1 \cdot t_2$ with $t_2 \in \mathbf{T}(U)$ and $t_1 \in \mathbf{T}(U^c)$, where $U^c = \{a_1, ..., a_n\} - U$, such that

$$t_2 = a_{i_1}^{\alpha_{i_1}} \cdots a_{i_r}^{\alpha_{i_r}}, \quad \alpha_{i_j} \geq M, \ 1 \leq j \leq r.$$

We must have $t_2 \in \mathcal{B}_{\leq m} - \langle \mathbf{LT}(L) \rangle$ because t_2 is a factor of t. On the other hand, since $r > d$, we have $\mathbf{T}(U) \cap \langle \mathbf{LT}(L) \rangle \neq \{0\}$, and so there exists $g \in \mathcal{G}$ with $\mathbf{LT}(g) \in \mathbf{T}(U)$. Note that since for $\mathbf{LT}(g) = c_\beta a_1^{\beta_1} \cdots a_n^{\beta_n}$ we have $\beta_i \leq M$, $1 \leq i \leq n$, it follows that t_2 is divisible by $\mathbf{LT}(g)$ and thus $t \in \langle \mathbf{LT}(L) \rangle$, a contradiction.

For the inequality "$\geq$", let $U = \{a_{i_1}, ..., a_{i_d}\} \subset \{a_1, ..., a_n\}$ be such that $i_1 > \cdots > i_d$ and $\mathbf{T}(U) \cap \langle \mathbf{LT}(L) \rangle = \{0\}$. Then for $m \geq n \cdot M$, it is easy to see that $t = a_{i_1}^M \cdots a_{i_d}^M \in \mathcal{B}_{\leq m} - \langle \mathbf{LT}(L) \rangle$ and $|\mathbf{top}_M(t)| = d$.   □

Furthermore, let $J = \sum_{j=1}^{s} Am_j$ be a left ideal of A generated by standard monomials (i.e., it is a monomial left ideal), where

$$m_j = a_1^{\alpha_{j1}} \cdots a_n^{\alpha_{jn}}, \quad j = 1, ..., s.$$

Then we may associate each m_j to the n-tuple of exponents integers

$$R_j = (\alpha_{j1}, \alpha_{j2}, ..., \alpha_{jn}) \in I\!N^n, \quad j = 1, ..., s,$$

and put

$$M_j = \Big\{ i \subset \{1, ..., n\} \Big| \alpha_{ji} \neq 0 \text{ in } R_j \Big\}, \quad j = 1, ..., s.$$

$$\mathcal{M} = \Big\{ J \subset \{1, ..., n\} \Big| J \cap M_j \neq \emptyset, \ 1 \leq j \leq s \Big\}.$$

From Proposition 1.4 and Proposition 2.1 we know that if L is a left ideal of A and $\mathcal{G}$ is a Gröbner basis of L, then $h_L(x) = h_{\langle \mathbf{LT}(L)]}(x) = h_{\langle \mathbf{LT}(\mathcal{G})]}(x)$. This fact enables us to derive the following algorithmic method for computing GK.dim(A/L).

4.2. Theorem With notation as before, let L be a left ideal of A and let $\mathcal{G} = \{g_1, ..., g_s\}$ be a left Gröbner basis of L. Put

$$m_j = \mathbf{LM}(g_j) = a_1^{\alpha_{j1}} \cdots a_n^{\alpha_{jn}}, \quad j = 1, ..., s.$$

Then

$$\text{GK.dim}(A/L) = d = n - \min\Big\{ |J| \ \Big| \ J \in \mathcal{M} \Big\}.$$

Consequently,
(i) $d = n$ if and only if $L = \{0\}$; and
(ii) if $\mathcal{G}$ is a reduced Gröbner basis (CH.II Definition 3.2(iii)), then $d = 0$ (i.e., A/L is a finite dimensional k-space) if and only if $s = n$ and (after reordering m_j's if necessary) $m_j = a_j^{r_j}$, $r_j > 0$, $j = 1, ..., n$. (Also see Theorem 8.1.)

Proof First note that for a subset $U \subset \{a_1, ..., a_n\}$, $T(U) \cap \langle \mathbf{LT}(L)] = \{0\}$ if and only if $T(U) \cap \mathbf{LT}(\mathcal{G}) = \emptyset$ where $\mathbf{LT}(\mathcal{G}) = \{\mathbf{LT}(g_1), ..., \mathbf{LT}(g_s)\}$. Now suppose $U = \{a_{i_1}, ..., a_{i_r}\} \subset \{a_1, ..., a_n\}$ with $i_1 > \cdots > i_r$ and $T(U) \cap \mathbf{LT}(\mathcal{G}) = \emptyset$. Then the set $J_U = \{1, ..., n\} - \{i_1, ..., i_r\}$ satisfies $J_U \cap M_j \neq \emptyset$, $1 \leq j \leq s$. Hence $J_U \in \mathcal{M}$, and consequently

$$r = n - |J_U| \leq n - \min\Big\{ |J| \ \Big| \ J \in \mathcal{M} \Big\}.$$

Conversely, for any $\{i_1, ..., i_t\} = J \subset \mathcal{M}$, if we put $U = \{a_{j_1}, ..., a_{j_{n-t}}\}$, where $\{j_1, ..., j_{n-t}\} = \{1, ..., n\} - J$, then it is easy to see that $T(U) \cap \mathbf{LT}(\mathcal{G}) = \emptyset$. Hence, $|U| = n - t \leq d$. Thus we have proved the requred equality. $\square$

Summing up, over a computable ground field k, one can proceed to compute GK.dim(A/L) as follows.

- Using a graded monomial ordering on $\mathcal{B}$ to compute a left Gröbner basis for L.
- Compute the number $d = \text{GK.dim}(A/L)$ by using Theorem 4.2. Note that our discussion has indeed given an algorithm for doing this.

We now go on to show that the independence condition for subsets of the generating set of A given above may be replaced by a weaker independence condition which will enable us to obtain an elimination lemma for homogeneous solvable polynomial algebras.

With notation as before, we say that a subset $U = \{a_{i_1}, ..., a_{i_r}\} \subset \{a_1, ..., a_n\}$ with $i_1 < i_2 < \cdots < i_r$ is *weakly independent* (modulo L) if $k[U] \cap L = \{0\}$. It is clear that if U is independent (modulo L), then U is weakly independent (modulo L). Hence, if we put

$$d' = \max\left\{ |U| \mid U \subset \{a_1, ..., a_n\} \text{ weakly independent (modulo } L)\right\},$$

it follows from Proposition 4.1 that

$(*)$ $\qquad\qquad\qquad\qquad\qquad d' \geq d = \deg h_L(x).$

4.3. Proposition With notation as before, we have

$$d' = d = \deg h_L(x).$$

Proof In view of the above $(*)$, we only have to show $d \geq d'$.
If $d' = 0$, then since L is a proper nonzero left ideal we have $h_L(m) \geq 1 = \begin{pmatrix} m+0 \\ 0 \end{pmatrix}$. So we suppose $d' > 0$, and without loss of generality we let $U = \{a_1, ..., a_{d'}\} \subset \{a_1, ..., a_n\}$ be weakly independent (modulo L). Consider the standard filtration $Fk[U]$ on $k[U]$ as we defined for A. Then $F_m k[U] \subset F_m A$ for all $m \in I\!N$. Since all standard monomials in variables $a_1, ..., a_{d'}$ form a k-basis for $k[U]$, it follows from the weak independence condition $k[U] \cap L = \{0\}$ that

$$\begin{pmatrix} m+d' \\ d' \end{pmatrix} = \dim_k F_m k[U] \quad = \quad \dim_k \frac{F_m k[U] + L}{L}$$

$$\leq \quad \dim_k \frac{F_m A + L}{L}$$

$$= \quad HF_L(m).$$

Thus for $m \gg 0$ we obtain

$$h_L(m) \geq \begin{pmatrix} m+d' \\ d' \end{pmatrix} = f(m),$$

where $f(x)$ denote the polynomial $\begin{pmatrix} x+d' \\ d' \end{pmatrix}$. Hence $d = \deg h_L(x) \geq \deg f(x) = d'$, as desired. $\qquad\qquad\square$

It follows immediately from Proposition 4.3 that the following lemma holds.

4.4. Lemma (elimination Lemma for homogeneous solvable polynomial algebras) Let $A = k[a_1, ..., a_n]$ be a homogeneous solvable polynomial algebra, and let L be a proper left ideal of A such that the A-module A/L has Gelfand-Kirillov dimension d, i.e., $\deg h_L(x) = d$. Then, for every subset $U = \{a_{i_1}, ..., a_{i_{d+1}}\} \subset \{a_1, ..., a_n\}$,

$$L \cap k[U] \neq \{0\},$$

where $k[U] = k[a_{i_1}, ..., a_{i_{d+1}}] \subset A$.

5. GK-dimension Computation and Elimination of Variables (Linear Case)

In this section we consider the GK-dimension computation of modules and the elimination of variables by restricting to the (left) admissible system $(A, \mathcal{B}, \succeq_{gr})$, where $A = k[a_1, ..., a_n]$ is a *linear* solvable polynomial algebra in the sense of CH.II Definition 2.1.

5.1. Proposition Let $A = k[a_1, ..., a_n]$ be a finitely generated k-algebra with the k-basis $\mathcal{B} = \{a^\alpha = a_1^{\alpha_1} a_2^{\alpha_2} \cdots a_n^{\alpha_n} \mid \alpha = (\alpha_1, \alpha_2, ..., \alpha_n) \in \mathbb{N}^n\}$. Consider the standard filtration FA. Then A is a linear solvable polynomial algebra with respect to some graded monomial ordering $\succeq_{gr}$ on $\mathcal{B}$ if and only if the associated graded k-algebra $G(A) = k[\sigma(a_1), ..., \sigma(a_n)]$ is a homogeneous solvable polynomial algebra with respect to $\succeq_{gr}$ on the k-basis $\sigma(\mathcal{B}) = \{\sigma(a^\alpha) \mid a^\alpha \in \mathcal{B}\}$.

Proof This easily follows from CH.IV Theorem 1.6 and Theorem 4.1. Or we refer to [LW] for a direct proof. □

5.2. Theorem Let $(A, \mathcal{B}, \succeq_{gr})$ be the (left) admissible system associated with a linear solvable polynomial algebra A, and let FA be the standard filtration on A. If L is a nonzero left ideal of A, then there exists a unique polynomial $h_L(x) \in \mathbb{Q}[x]$ with positive leading coefficient such that for $m \gg 0$,

$$\dim_k \left(\frac{F_m A}{F_m L} \right) = h_L(m),$$

where $F_m L = F_m A \cap L$. Moreover, if the ground field is computable, then the polynomial $h_L(x)$ can be computed from any given generating set of L.

Proof Note that $F_m A / F_m L \cong (F_m A + L)/L$, and the latter is the filtration on A/L induced by FA. It follows from CH.I §3 that we have $G(A/L) \cong$

$G(A)/G(L)$, and hence

$$\dim_k \left(\frac{F_m A}{F_m L} \right) = \dim_k \left(\frac{\oplus_{p \leq m} G(A)_p + G(L)}{G(L)} \right), \quad m \in I\!N.$$

Note that $\oplus_{p \leq m} G(A)_p = F_m G(A)$, the mth part of the standard filtration $FG(A)$ on $G(A)$.

On the other hand, Proposition 5.1 entails that $G(A)$ is a homogeneous solvable polynomial k-algebra with $\succeq_{gr}$. Now the existence of the required polynomial $h_L(x)$ follows from Theorem 3.3.

If furthermore k is computable and $L = \sum_{i=1}^{s} A\xi_i$, $\xi_i \in L$, then we can compute $h_L(x)$ as follows.

- Compute a Gröbner basis $\mathcal{G} = \{g_1, ..., g_s\}$ of L starting with $\{\xi_1, ..., \xi_s\}$.
- Take $\sigma(\mathcal{G}) = \{\sigma(g_1), ..., \sigma(g_s)\}$ as a Gröbner basis of $G(L)$ by CH.IV Theorem 2.1.
- Proceed from here as in the case of Theorem 3.3.

□

5.3. Definition The polynomial obtained in Theorem 5.2 is called the *Hilbert polynomial* of the A-module A/L and is denoted by $h_L(x)$.

The above argumentation has indeed shown that the degree of the Hilbert polynomial of the A-module A/L is nothing but the Gelfand-Kirillov dimension of A/L, i.e.,

$$(\nabla) \qquad \begin{aligned} \deg h_L(x) = \text{GK.dim}(A/L) &= \text{GK.dim}(G(A/L)) \\ &= \text{GK.dim}(G(A)/G(L)) \\ &= \deg h_{G(L)}(x). \end{aligned}$$

Hence, by §4, we immediately have the following

5.4. Corollary Let A and L be as in Theorem 5.2. If the ground field k is computable, then there is an algorithm for computing $\text{GK.dim}(A/L)$ from any given generating set of L.

□

Example (i) Let $A = A_1(q)$ be the additive analogue of the first Weyl algebra over a field k (CH.I §5), i.e., $A = k[x, y]$ subject to the relation $yx - qxy = 1$, where q is a nonzero element of k. Then A is a linear solvable polynomial algebra with respect to $x >_{grlex} y$. Let $L = Af + Ag$, where $f = qx$, $g = xy + q^{-1}$. By a direct calculation, the S-element $S(f, g)$ of f and g is 0. Hence, $\mathcal{G} = \{f, g\}$ is a left Gröbner basis of L. Since $\mathbf{LM}(f) = x$, $\mathbf{LM}(g) = xy$, it follows from the procedure of §3 that the Hilbert polynomial of A/L is $h_L(x) = x \in \mathbb{Q}[x]$. Thus, $\text{GK.dim}(A/L) = 1$ and A/L is an infinite dimensional A-module.

We now proceed to determine the existence of an elimination lemma for a linear solvable polynomial algebra by computing GK.dim(A/L) directly without computing $h_L(x)$.

Let $A = k[a_1, ..., a_n]$ be a linear solvable polynomial algebra with the standard filtration FA. Note that the relations satisfied by the generators of A (CH.III §2) entails that, for a nonempty subset $U = \{a_{i_1}, ..., a_{i_r}\}$ of $\{a_1, ..., a_n\}$, the subalgebra $k[U]$ of A generated by U over k may contain elements which are not the linear sum of the standard monomials in variables in U. Hence, the method we used for the homogeneous case in §4 cannot be carried over to the linear solvable case. However, since $G(A)$ is a homogeneous solvable polynomial algebra (Proposition 5.1), by using the filtered-graded transfer trick again, we may still arrive at an elimination lemma for linear solvable polynomial algebras.

Let L be a left ideal of A with the filtration FL induced by the standard filtration FA on A, and let $\mathcal{G} = \{g_1, ..., g_s\}$ be a left Gröbner basis of L. It follows from CH.IV Theorem 2.1 that $\sigma(\mathcal{G}) = \{\sigma(g_1), ..., \sigma(g_s)\}$ is a left Gröbner basis of $G(L)$ in $G(A)$. Writing

$$\begin{cases} m_j = \mathbf{LM}(g_j) = a_1^{\alpha_{j1}} \cdots a_n^{\alpha_{jn}}, \\ R_j = (\alpha_{j1}, ..., \alpha_{jn}), \end{cases} \quad j = 1, ..., s,$$

then CH.IV Lemma 1.2 and Theorem 4.1 yield

$$(\Delta) \qquad \begin{aligned} \mathbf{LM}\left(\sigma(g_j)\right) &= \sigma\left(\mathbf{LM}(g_j)\right) \\ &= \sigma(a_1)^{\alpha_{j1}} \cdots \sigma(a_n)^{\alpha_{jn}}. \end{aligned}$$

Using notation as above and as in §4:

$$M_j = \left\{ i \in \{1, ..., n\} \mid \alpha_{ji} \neq 0 \text{ in } R_j \right\}, \ j = 1, ..., s,$$

$$\mathcal{M} = \left\{ J \subset \{1, ..., n\} \mid J \cap M_j \neq \emptyset, \ 1 \leq j \leq s \right\},$$

we may also compute GK.dim(A/L) as follows.

5.5. Proposition GK.dim$(A/L) = n - \min\left\{ |J| \mid J \in \mathcal{M} \right\}$. Consequently,
(i) $d = n$ if and only if $L = \{0\}$; and
(ii) if $\mathcal{G}$ is a reduced Gröbner basis (CH.II Definition 3.2(iii)), then $d = 0$ (i.e., A/L is a finite dimensional k-space) if and only if $s = n$ and (after reordering m_j's if necessary) $m_j = a_j^{r_j}$, $r_j > 0$, $j = 1, ..., n$. (Also see Theorem 8.1 in this chapter.)

Proof This follows immediately from the foregoing (∇), (Δ), and Theorem 4.2. $\qquad \square$

Now, for a nonempty subset $U = \{a_{i_1}, ..., a_{i_r}\} \subset \{a_1, ..., a_n\}$ with $i_1 < i_2 < \cdots < i_r$, we write

$$\mathbf{T}(U) = \left\{ \lambda a_{i_1}^{\alpha_1} \cdots a_{i_r}^{\alpha_r} \;\middle|\; \lambda \in k, \; (\alpha_1, ..., \alpha_r) \in I\!N^r \right\},$$

and we say that U is *independent* (modulo L) if $\mathbf{T}(U) \cap \mathbf{LT}(\mathcal{G}) = \emptyset$, where $\mathbf{LT}(\mathcal{G}) = \{\mathbf{LT}(g_1), ..., \mathbf{LT}(g_s)\}$.

5.6. Proposition With notation as above, and putting

$$d' = \max \left\{ |U| \;\middle|\; U \subset \{a_1, ..., a_n\} \text{ independent (modulo } L) \right\},$$

then $\mathrm{GK.dim}(A/L) = d'$.

Proof let $J \in \mathcal{M}$ with $J = \{i_1, ..., i_r\} \subset \{1, ..., n\}$. Then $J \cap M_j \neq \emptyset, j = 1, ..., s$. Write $U = \{a_{j_1}, ..., a_{j_{n-r}}\}$ with $\{j_1, ..., j_{n-r}\} = \{1, ..., n\} - J$. It is easy to see that $\mathbf{T}(U) \cap \mathbf{LT}(\mathcal{G}) = \emptyset$, i.e., $n - r = |U| \leq d'$. Since J is arbitrary in $\mathcal{M}$ it follows from Proposition 5.5 that $\mathrm{GK.dim}(A/L) \leq d'$.

To obtain the opposite inequality, let $U = \{a_{i_1}, ..., a_{i_{d'}}\} \subset \{a_1, ..., a_n\}$ be independent (modulo L), and put $J = \{1, ..., n\} - \{i_1, ..., i_{d'}\}$. Then we see that $J \in \mathcal{M}$. Thus, $d' = |U| = n - |J| \leq \mathrm{GK.dim}(A/L)$ by Proposition 5.5. This proves the proposition. $\square$

Furthermore, let $U = \{a_{i_1}, ..., a_{i_r}\} \subset \{a_1, ..., a_n\}$ with $i_1 < i_2 < \cdots < i_r$, and let

$$\mathcal{B}(U) = \left\{ a_{i_1}^{\alpha_1} \cdots a_{i_r}^{\alpha_r} \;\middle|\; (\alpha_1, ..., \alpha_r) \in I\!N^r \right\}.$$

Write $\mathbf{V}(U)$ for the *k-vector space* spanned by $\mathcal{B}(U)$ in A. We say that U is *weakly independent* (modulo L), if $\mathbf{V}(U) \cap L = \{0\}$.

5.7. Theorem With notation as above, if we put

$$d'' = \max \left\{ |U| \;\middle|\; U \subset \{a_1, ..., a_n\} \text{ weakly independent (modulo } L) \right\},$$

then $\mathrm{GK.dim}(A/L) = d''$.

Proof We first prove that if $U = \{a_{i_1}, ..., a_{i_r}\} \subset \{a_1, ..., a_n\}$ is independent (modulo L) with respect to some fixed Gröbner basis $\mathcal{G}$ of L, then U is weakly independent (modulo L), and hence $\mathrm{GK.dim}(A/L) \leq d''$ by Proposition 5.6. To see this, let $\mathcal{G} = \{g_1, ..., g_s\}$, and suppose that $f \in \mathbf{V}(U) \cap L$. If $f \neq 0$, then f has a Gröbner presentation by $\mathcal{G}$:

$$f = \sum_{i=1}^{s} h_i g_i, \quad \mathbf{LM}(f) \succeq_{gr} \mathbf{LM}(h_i g_i) \text{ whenever } h_i g_i \neq 0.$$

Hence, $\mathbf{LM}(f)$ appears as one of the $\mathbf{LM}(\mathbf{LT}(h_i)\mathbf{LT}(g_i))$. But $\mathbf{LM}(f) \in \mathcal{B}(U)$, this implies that some $\mathbf{LT}(g_i)$ must be contained in $\mathbf{T}(U) \cap \mathbf{LT}(\mathcal{G})$, contradicting U being independent (modulo L).

To prove $d'' \leq \mathrm{GK.dim}(A/L)$, let $h_L(x)$ be the Hilbert polynomial of the A-module A/L. If $d'' = 0$, then since L is a proper left ideal we have $h_L(m) \geq 1 = \begin{pmatrix} m+0 \\ 0 \end{pmatrix}$. So we may suppose $d'' > 0$, and without loss of generality we let $U = \{a_1, ..., a_{d''}\} \subset \{a_1, ..., a_n\}$ be weakly independent (modulo L). Consider the filtration $F\mathbf{V}(U)$ on the k-vector space $\mathbf{V}(U)$ induced by FA: $F_m\mathbf{V}(U) = F_mA \cap \mathbf{V}(U)$, $m \subset I\!N$, and put

$$\mathcal{B}(U)_{\leq m} = \left\{ a_1^{\alpha_1} \cdots a_{d''}^{\alpha_{d''}} \ \middle|\ \alpha_1 + \cdots + \alpha_{d''} \leq m \right\}, \ m \in I\!N.$$

Then $\mathcal{B}(U)_{\leq m} \subset F_m\mathbf{V}(U)$, and it follows from the weak independence condition $\mathbf{V}(U) \cap L = \{0\}$ that for $m \gg 0$

$$\begin{pmatrix} m+d'' \\ d'' \end{pmatrix} = |\mathcal{B}(U)_{\leq m}| \ = \ \dim_k \frac{F_m\mathbf{V}(U) + L}{L}$$

$$\leq \ \dim_k \frac{F_mA + L}{L}$$

$$= \ h_L(m).$$

Thus we obtain

$$h_L(m) \geq \begin{pmatrix} m+d'' \\ d'' \end{pmatrix} = f(m), \ m \gg 0,$$

where $f(x)$ denote the polynomial $\begin{pmatrix} x+d'' \\ d'' \end{pmatrix}$. Hence $\mathrm{GK.dim}(A/L) = \deg h_L(x) \geq \deg f(x) = d''$, as desired. $\qquad\qquad\square$

We arrive at the following lemma.

5.8. Lemma (elimination Lemma for linear solvable polynomial algebras) Let $A = k[a_1, ..., a_n]$ be a linear solvable polynomial algebra with the associated (left) admissible system $(A, \mathcal{B}, \succeq_{gr})$, and let L be a proper left ideal of A such that the A-module A/L has Gelfand-Kirillov dimension d, i.e., the Hilbert polynomial of the A-module A/L has degree d. Then for every subset $U = \{a_{i_1}, ..., a_{i_{d+1}}\} \subset \{a_1, ..., a_n\}$ with $i_1 < i_2 < \cdots < i_{d+1}$,

$$\mathbf{V}(U) \cap L \neq \{0\},$$

where $\mathbf{V}(U) = k\text{-Span} \left\{ a_{i_1}^{\alpha_1} \cdots a_{i_{d+1}}^{\alpha_{d+1}} \ \middle|\ (\alpha_1, ..., \alpha_{d+1}) \in I\!N^{d+1} \right\}.$

6. The $\succeq_{gr}$-filtration on a Quadric Solvable Polynomial Algebra

Let $A = k[a_1, ..., a_n]$ be a quadric solvable polynomial algebra with the associated (left) admissible system $(A, \mathcal{B}, \succeq_{gr})$, where

$$\mathcal{B} = \left\{ a^\alpha = a_1^{\alpha_1} \cdots a_n^{\alpha_n} \ \middle| \ \alpha = (\alpha_1, ..., \alpha_n) \in I\!N^n \right\}$$

is the standard k basis and $\succeq_{gr}$ is a graded monomial ordering on $\mathcal{B}$. Recall from CH.III Definition 2.1 that the generators of A satisfy

$$a_j a_i = \lambda_{ji} a_i a_j + \sum_{k \le \ell} \lambda_{ji}^{k\ell} a_k a_\ell + \sum \lambda_h a_h + c_{ji}, \ 1 \le i < j \le n.$$

Suppose that A is *neither* homogeneous *nor* linear. Then it is clear that, with respect to the standard filtration FA on A, $G(A)$ is *no longer* a homogeneous solvable polynomial algebra (though $G(A)$ is a solvable polynomial algebra by CH.IV Theorem 4.1). It follows that the method we used in computing GK-dimension of modules over linear solvable polynomial algebras in §5 cannot be directly carried over to *arbitrary* quadric solvable polynomial algebras. To remedy this problem, we first introduce the $\succeq_{gr}$-filtration $\mathcal{F}A$ on A in this section.

With notation as above, for each $\alpha \in I\!N^n$, construct the k-subspace

$$\mathcal{F}_\alpha A = k\text{-span} \left\{ a^\beta \in \mathcal{B} \ \middle| \ \alpha \succeq_{gr} \beta \right\}.$$

Clearly, if $\alpha \succ_{gr} \gamma$, then $\mathcal{F}_\gamma A \subset \mathcal{F}_\alpha A$. Thus, since $\succeq_{gr}$ is a monomial ordering, we have a $I\!N^n$-filtration on A satisfying
(1) $1 \in \mathcal{F}_0 A$,
(2) every $\mathcal{F}_\alpha A$ is a *finite* dimensional k-space, $A = \cup_{\alpha \in I\!N^n} \mathcal{F}_\alpha A$, and
(3) $(\mathcal{F}_\alpha A)(\mathcal{F}_\beta A) \subset \mathcal{F}_{\alpha + \beta} A$.
To emphasise the role of $\succeq_{gr}$ in our discussion, we prefer calling this filtration $\mathcal{F}A$ the $\succeq_{gr}$-*filtration* instead of simply a $I\!N^n$-filtration as it looks like.
Since $\succeq_{gr}$ is a monomial ordering, it is known that $\alpha \succeq_{gr} 0 = (0, ..., 0)$ for all $\alpha \in I\!N^n$; and since $\succeq_{gr}$ is a graded monomial ordering, for each $\alpha \in I\!N^n$, there exists

$$\alpha^* = \max \left\{ \gamma \in I\!N^n \ \middle| \ \alpha \succ_{gr} \gamma \right\}.$$

Then we have a well-defined $I\!N^n$-graded algebra

$$G^{\mathcal{F}}(A) = \bigoplus_{\alpha \in I\!N^n} G^{\mathcal{F}}(A)_\alpha \text{ with } G^{\mathcal{F}}(A)_\alpha = \frac{\mathcal{F}_\alpha A}{\mathcal{F}_{\alpha^*} A},$$

in which the addition is given by the componentwise addition and the multiplication is given by

$$G^{\mathcal{F}}(A)_\alpha \times G^{\mathcal{F}}(A)_\beta \longrightarrow G^{\mathcal{F}}(A)_{\alpha + \beta}$$

$$(\overline{f}, \overline{g}) \longmapsto \overline{fg}$$

where, if $f \in \mathcal{F}_\alpha A$, then $\overline{f}$ stands for the image of f in $G^{\mathcal{F}}(A)_\alpha = \mathcal{F}_\alpha A / \mathcal{F}_{\alpha^\bullet} A$. $G^{\mathcal{F}}(A)$ is called the *associated graded algebra* of A with respect to $\mathcal{F}A$.
As dealing with a $\mathbb{Z}$-filtration in CH.I §3, for an element $f \in \mathcal{F}_\alpha A - F_{\alpha^\bullet} A$, we write $\sigma(f)$ for the image of f in $G^{\mathcal{F}}(A)_\alpha$. For each $i = 1, ..., n$, write e_i for the ith unit vector $(0, ..., 0, 1, 0, ..., 0)$ in $\mathbb{N}^n$. Then $0 \neq \sigma(a_i) \in G^{\mathcal{F}}(A)_{e_i}$, and it is not hard to see that, for $\alpha = (\alpha_1, ..., \alpha_n) \in \mathbb{N}^n$ and $a^\alpha \in \mathcal{B}$,

$$\sigma(a_1)^{\alpha_1} \cdots \sigma(a_n)^{\alpha_n} = \sigma(a_1^{\alpha_1} \cdots a_n^{\alpha_n}) = \sigma(a^\alpha).$$

Hence, for $\alpha = (\alpha_1, ..., \alpha_n) \in \mathbb{N}^n$,

$$G^{\mathcal{F}}(A)_\alpha = k\text{-span}\{\sigma(a_1)^{\alpha_1} \cdots \sigma(a_n)^{\alpha_n}\} \text{ (i.e., a 1-dimensional space)},$$

and consequently, we have indeed proved the following result.

6.1. Proposition With notation as above, the following holds.
(i) $G^{\mathcal{F}}(A) = k[\sigma(a_1), ..., \sigma(a_n)]$ is a $\mathbb{N}^n$-graded k-algebra generated by $\sigma(a_1)$,..., $\sigma(a_n)$. The generators of $G^{\mathcal{F}}(A)$ satisfy

$$\sigma(a_j)\sigma(a_i) - \lambda_{ij}\sigma(a_i)\sigma(a_j), \quad j > i, \quad \lambda_{ij} \neq 0.$$

(ii) The set of monomials

$$\sigma(\mathcal{B}) = \left\{ \sigma(a_1)^{\alpha_1} \cdots \sigma(a_n)^{\alpha_n} \mid (\alpha_1, ..., \alpha_n) \in \mathbb{N}^n \right\}$$

forms a k-basis for $G^{\mathcal{F}}(A)$.
It follows that $G^{\mathcal{F}}(A)$ forms a homogeneous solvable polynomial algebra (in the sense of CH.III Definition 2.1) with respect to the monomial ordering $\succeq_{gr}$. And it follows that $G^{\mathcal{F}}(A)$ is a Noetherian domain.

$\square$

For each integer $p \in \mathbb{N}$, we now define the *p-block* of $\mathbb{N}^n$ as

$$\Gamma_p = \left\{ \alpha \in \mathbb{N}^n \mid |\alpha| = p \right\},$$

where if $\alpha = (\alpha_1, ..., \alpha_n)$, then $|\alpha| = \alpha_1 + \cdots + \alpha_n$. Writing $s(p) = |\Gamma_p|$, then

$$s(p) = \binom{n + p - 1}{p}.$$

Hence, we can write

$$\Gamma_p = \{\alpha(1), \alpha(2), ..., \alpha(s(p) - 1), \alpha(s(p))\},$$

and without loss of generality, we may assume that

$$\alpha(s(p)) \succ_{gr} \alpha(s(p) - 1) \succ_{gr} \cdots \succ_{gr} \alpha(2) \succ_{gr} \alpha(1).$$

Returning to the standard filtration FA on A, the next lemma makes the *key* link between $\mathcal{F}A$ and FA.

6.2. Lemma With notation as above, we have

$$\mathcal{F}_{\alpha(s(p))}A = F_p A, \text{ for all } p \in I\!N.$$

Proof Since $\succeq_{gr}$ is a graded monomial ordering, using the p-block defined above, this follows from a careful comparison of $\mathcal{F}_{\alpha(s(p))}A$ and $F_p A$.

7. GK-dimension Computation and Elimination of Variables (General Quadric Case)

let $A = k[a_1, ..., a_n]$ be an arbitrary quadric solvable polynomial algebra with the associated (left) admissible system $(A, \mathcal{B}, \succeq_{gr})$, and let $\mathcal{F}A$ be the $\succeq_{gr}$-filtration on A (as defined in §6). For a given left ideal $L \subset A$, we now start to compute the Gelfand-Kirillov dimension GK.dim$M$ of the left A-module $M = A/L$ by using a (left) Gröbner basis of L, and then derive an elimination lemma for A. All notation are retained from previous sections.

7.1. Lemma (i) Let $f, g \in A$. Then $\sigma(f)\sigma(g) = \sigma(fg)$ in $G^{\mathcal{F}}(A)$.
(ii) If $f = \sum_{i=1}^{s} c_i a^{\alpha(i)}$, where $a^{\alpha(i)} \in \mathcal{B}$ and $0 \neq c_i \in k$, is such that

$$\mathbf{LM}(f) = a_1^{\alpha_1} \cdots a_n^{\alpha_n} = a^{\alpha(1)} \succ_{gr} a^{\alpha(2)} \succ_{gr} \cdots \succ_{gr} a^{\alpha(s)},$$

Then $\mathbf{LM}(\sigma(f)) = \sigma(\mathbf{LM}(f)) = \sigma(a_1)^{\alpha_1} \cdots \sigma(a_n)^{\alpha_n} = \sigma(f) \in G^{\mathcal{F}}(A)_{\alpha(1)}$.

Proof Since $G^{\mathcal{F}}(A)$ is a homogeneous solvable polynomial algebra with respect to $\succeq_{gr}$ (Proposition 6.1), (i) and (ii) are then clear by the definition of $\mathcal{F}A$, the definition of $\mathbf{LM}(f)$, and the definition of $\sigma(f)$.   □

7.2. Proposition Let L be a left ideal of A and $\mathcal{G} = \{g_1, ..., g_s\}$ a Gröbner basis for L. Then

$$\sigma(\mathcal{G}) = \{\sigma(g_1), ..., \sigma(g_s)\} = \{\sigma(\mathbf{LM}(g_1)), ..., \sigma(\mathbf{LM}(g_s))\}$$

and $\sigma(\mathcal{G})$ forms a Gröbner basis for $G^{\mathcal{F}}(L) \subset G^{\mathcal{F}}(A)$, where $G^{\mathcal{F}}(L)$ is the associated graded left ideal of L with respect to the induced filtration: $\mathcal{F}_\alpha L = L \cap \mathcal{F}_\alpha A$, $\alpha \in I\!N^n$.

Proof The fact that $\sigma(\mathcal{G}) = \{\sigma(g_1) = \sigma(\mathbf{LM}(g_1)), ..., \sigma(g_s) = \sigma(\mathbf{LM}(g_s))\}$ follows from Lemma 7.1. Since $G^{\mathcal{F}}(A)$ is a homogeneous solvable polynomial algebra by Proposition 6.1, it follows from CH.V §1 that $\sigma(\mathcal{G})$ forms a left Gröbner basis in $G^{\mathcal{F}}(A)$. To see that $\sigma(\mathcal{G})$ is a Gröbner basis for $G^{\mathcal{F}}(L)$ in $G^{\mathcal{F}}(A)$, we need to prove that every homogeneous element of $G^{\mathcal{F}}(L)$ has a Gröbner presentation by $\sigma(\mathcal{G})$.

Let $f, g, h \in A$ be such that $\mathbf{LM}(f) \succeq_{gr} \mathbf{LM}(hg)$. Then it follows from Proposition 6.1 and Lemma 7.1 that

$$
\begin{aligned}
\mathbf{LM}(\sigma(f)) &= \sigma(\mathbf{LM}(f)) \\
&\succeq_{gr} \sigma(\mathbf{LM}(hg)) \\
&= \mathbf{LM}(\sigma(hg)) = \mathbf{LM}(\sigma(h)\sigma(g)).
\end{aligned}
$$

Now, if $f \in L$ and $f = \sum_{i=1}^{s} h_i g_i$ is a Gröbner presentation of f by $\mathcal{G}$, then $\mathbf{LM}(f) \succeq_{gr} \mathbf{LM}(h_i g_i)$ whenever $h_i g_i \neq 0$. Thus, $\sigma(f) = \sum \sigma(h_i g_i) = \sum \sigma(h_i)\sigma(g_i)$ with the property that $\mathbf{LM}(\sigma(f)) \succeq_{gr} \mathbf{LM}(\sigma(h_i)\sigma(g_i))$ whenever $\sigma(h_i g_i) \neq 0$. This shows that every homogeneous element of $G^{\mathcal{F}}(L)$ has a Gröbner presentation by $\sigma(\mathcal{G})$, and hence, $\sigma(\mathcal{G})$ is a Gröbner basis for $G^{\mathcal{F}}(L)$ in $G^{\mathcal{F}}(A)$. $\qquad\square$

7.3. Proposition Let L be as in Proposition 7.2 and $M = A/L$. Then

$$
\mathrm{GK.dim}M = \mathrm{GK.dim}\frac{G^{\mathcal{F}}(A)}{G^{\mathcal{F}}(L)}
$$

Proof In order to compute GK.dim$M$, our idea is to use the relation between the $\succeq_{gr}$-*filtration* $\mathcal{F}A$ and the *standard filtration* FA (see Lemma 6.2):

$$
(1) \quad
\begin{cases}
\mathcal{F}_{\alpha(s(p))}A = F_p A, \text{ and hence} \\
\dim_k \mathcal{F}_{\alpha(s(p))}A = \dim_k F_p A = \dbinom{n+p}{p}, \; p \in \mathbb{N}.
\end{cases}
$$

To better understand the proof given below, we refer the reader to CH.VIII §2. Let $\mathcal{F}M$ be the $\succeq_{gr}$-filtration on M induced by $\mathcal{F}A$: $\mathcal{F}_\alpha M = (\mathcal{F}_\alpha A + L)/L$, $\alpha \in \mathbb{N}^n$. Then M forms a $\succeq_{gr}$-filtered A-module in the sense that $\mathcal{F}_\alpha A \cdot \mathcal{F}_\beta M \subset \mathcal{F}_{\alpha+\beta}M$ for all $\alpha, \beta \in \mathbb{N}^n$. Let FM be the filtration on M induced by FA: $F_p M = (F_p A + L)/L$, $p \in \mathbb{N}$. Then (1) yields

$$
(2) \quad \mathcal{F}_{\alpha(s(p))}M = F_p M, \; p \in \mathbb{N}.
$$

Consider the associated graded $G^{\mathcal{F}}(A)$-module $G^{\mathcal{F}}(M) = \oplus_{\alpha \in \mathbb{N}^n} G^{\mathcal{F}}(M)_\alpha$, where $G^{\mathcal{F}}(M)_\alpha = \mathcal{F}_\alpha M/\mathcal{F}_{\alpha^{\bullet}} M$ and the module action of $G^{\mathcal{F}}(A)$ on $G^{\mathcal{F}}(M)$ is the action induced by the action of A on M. If we use the filtration $\mathcal{F}L$ on L induced by $\mathcal{F}A$, then there is the isomorphism of $\mathbb{N}^n$-graded $G^{\mathcal{F}}(A)$-modules

$$
(3) \quad
\begin{cases}
G^{\mathcal{F}}(M) = G^{\mathcal{F}}(A/L) \cong \dfrac{G^{\mathcal{F}}(A)}{G^{\mathcal{F}}(L)} = \displaystyle\bigoplus_{\alpha \in \mathbb{N}^n} \dfrac{G^{\mathcal{F}}(A)_\alpha + G^{\mathcal{F}}(L)}{G^{\mathcal{F}}(L)} \text{ with} \\
G^{\mathcal{F}}(M)_\alpha \cong \dfrac{G^{\mathcal{F}}(A)_\alpha + G^{\mathcal{F}}(L)}{G^{\mathcal{F}}(L)}, \; \alpha \in \mathbb{N}^n
\end{cases}
$$

After a finitely many seccessive use of the exact sequence of k-vector spaces

$$0 \to \mathcal{F}_{\alpha^*} M \longrightarrow \mathcal{F}_\alpha M \longrightarrow G^{\mathcal{F}}(M)_\alpha \to 0$$

we have

(4)
$$\begin{cases} \dim_k \mathcal{F}_{\alpha(s(p))} M = \sum_{\alpha(s(p)) \succeq_{gr} \alpha} \dim_k G^{\mathcal{F}}(M)_\alpha \\[2mm] \qquad\qquad = \dim_k \left(\bigoplus_{\alpha(s(p)) \succeq_{gr} \alpha} G^{\mathcal{F}}(M)_\alpha \right). \end{cases}$$

If we define

(5)
$$F_p \left(\frac{G^{\mathcal{F}}(A)}{G^{\mathcal{F}}(L)} \right) \doteq \bigoplus_{\alpha(s(p)) \succeq_{gr} \alpha} \frac{G^{\mathcal{F}}(A)_\alpha + G^{\mathcal{F}}(L)}{G^{\mathcal{F}}(L)}, \quad p \in \mathbb{N},$$

then this is a $\mathbb{N}$-filtration on $G^{\mathcal{F}}(A)/G^{\mathcal{F}}(L)$ consisting of finite dimensional k-spaces. With this filtration, $G^{\mathcal{F}}(A)/G^{\mathcal{F}}(L)$ forms a filtered $G^{\mathcal{F}}(A)$-module with respect to the *standard* filtration $FG^{\mathcal{F}}(A)$ on $G^{\mathcal{F}}(A)$, i.e., by definition,

(6)
$$\begin{cases} F_q G^{\mathcal{F}}(A) = \bigoplus_{\alpha(s(q)) \succeq_{gr} \beta} G^{\mathcal{F}}(A)_\beta, \quad q \in \mathbb{N}, \text{ and} \\[3mm] F_q G^{\mathcal{F}}(A) \cdot F_p \left(\frac{G^{\mathcal{F}}(A)}{G^{\mathcal{F}}(L)} \right) \subset F_{q+p} \left(\frac{G^{\mathcal{F}}(A)}{G^{\mathcal{F}}(L)} \right), \quad p, q \in \mathbb{N}. \end{cases}$$

(Indeed, one may see that the filtration we defined on $G^{\mathcal{F}}(A)/G^{\mathcal{F}}(L)$ above is nothing but the filtration induced by the standard filtration $FG^{\mathcal{F}}(A)$: $F_p(G^{\mathcal{F}}(A)/G^{\mathcal{F}}(L)) = (F_p G^{\mathcal{F}}(A) + G^{\mathcal{F}}(L))/G^{\mathcal{F}}(L), \; p \in \mathbb{N}.$) Consequently, the above $(2)+(3)+(4)+(5)+(6)$ yields

$$\dim_k F_p M = \dim_k \mathcal{F}_{\alpha(s(p))} M = \dim_k F_p \left(\frac{G^{\mathcal{F}}(A)}{G^{\mathcal{F}}(L)} \right), \quad p \in \mathbb{N},$$

and therefore, as filtered modules over the filtered algebra A with the standard filtration FA and the filtered algebra $G^{\mathcal{F}}(A)$ with the standard filtration $FG^{\mathcal{F}}(A)$, respectively, M and $G^{\mathcal{F}}(A)/G^{\mathcal{F}}(L)$ have the same growth measured by the same dimension function. Consequently, GK.dim$M$ = GK.dim$(G^{\mathcal{F}}(A)/G^{\mathcal{F}}(L))$. $\square$

Thus, the computation of GK.dim$M$ has been translated into the computation of GK.dim$(G^{\mathcal{F}}(A)/G^{\mathcal{F}}(L))$ while $G^{\mathcal{F}}(A)/G^{\mathcal{F}}(L)$ is a left graded module over the homogeneous solvable polynomial algebra $G^{\mathcal{F}}(A)$. However, note that if $\mathcal{G} = \{g_1, ..., g_s\}$ is the Gröbner basis for L as given in Proposition 7.2, then by Lemma 7.1, for $\mathbf{LM}(\mathcal{G}) = \{\mathbf{LM}(g_j) = a_1^{\alpha_{j1}} \cdots a_n^{\alpha_{jn}} \mid j = 1, ..., s\}$,

$$\sigma(\mathcal{G}) = \left\{ \sigma(\mathbf{LM}(g_j)) = \sigma(a_1)^{\alpha_{j1}} \cdots \sigma(a_n)^{\alpha_{jn}} \; \middle| \; j = 1, ..., s \right\}$$

is a Gröbner basis for $G^{\mathcal{F}}(L)$ in $G^{\mathcal{F}}(A)$. It follows from the results of §§3–4 for homogeneous solvable polynomial algebras that the computation of GK.dim$M$ can be realized via $\mathbf{LM}(\mathcal{G})$ as follows.

7.4. Theorem With notation as above, the following holds.
(i) There exists a unique polynomial $h_L(x)$ of rational coefficients with positive leading coefficient such that for $m \gg 0$,

$$\dim_k(F_p M) = h_L(m).$$

Hence, GK.dim$M = \deg h_L(x)$. Moreover, if the ground field is computable, then the polynomial $h_L(x)$ can be computed as in §3.
(ii) For $\mathbf{LM}(\mathcal{G}) = \{\mathbf{LM}(g_j) = a_1^{\alpha_{j1}} \cdots a_n^{\alpha_{jn}} \mid j = 1, ..., s\}$, put

$$R_j = (\alpha_{j1}, ..., \alpha_{jn}), \ j = 1, ..., s,$$

$$M_j = \Big\{ i \in \{1, ..., n\} \ \Big| \ \alpha_{ji} \neq 0 \text{ in } R_j \Big\}, \ j = 1, ..., s,$$

$$M = \Big\{ J \subset \{1, ..., n\} \ \Big| \ J \cap M_j \neq \emptyset, \ 1 \leq j \leq s \Big\}.$$

Then we have

$$\text{GK.dim}M = n - \min \Big\{ |J| \ \Big| \ J \in M \Big\}.$$

Consequently,
(a) $d = n$ if and only if $L = \{0\}$; and
(b) if $\mathcal{G}$ is a reduced Gröbner basis (CH.II Definition 3.2(iii)), then $d = 0$ (i.e., A/L is a finite dimensional k-space) if and only if $s = n$ and (after reordering m_j's if necessary) $m_j = a_j^{r_j}$, $r_j > 0$, $j = 1, ..., n$. (Also see Theorem 8.1 in next section.)

$\square$

As before, we call the polynomial $h_L(x)$ obtained in Theorem 7.4(i) the *Hilbert polynomial* of the module $M = A/L$.

Eventually, the above theorem enables us to use exactly the same argumentation as in the proof of Proposition 5.6 and Theorem 5.7 for obtaining the following lemma.

7.5. Lemma (elimination lemma for quadric solvable polynomial algebras) Let $A = k[a_1, ..., a_n]$ be a quadric solvable polynomial algebra with the associated (left) admissible system $(A, \mathcal{B}, \succeq_{gr})$, and let L be a proper left ideal of A such that the A-module A/L has Gelfand-Kirillov dimension d, i.e., the Hilbert polynomial of the A-module A/L has degree d. Then for every subset $U = \{a_{i_1}, ..., a_{i_{d+1}}\} \subset \{a_1, ..., a_n\}$ with $i_1 < i_2 < \cdots < i_{d+1}$,

$$\mathbf{V}(U) \cap L \neq \{0\},$$

where $\mathbf{V}(U) = k\text{-span}\{a_{i_1}^{\alpha_1} \cdots a_{i_{d+1}}^{\alpha_{d+1}} \mid (\alpha_1, ..., \alpha_{d+1}) \in I\!N^{d+1}\}.$

$\square$

Remark (i) In the commutative case, the notion of (weak) independence (modulo a polynomial ideal $I \subset k[x_1, ..., x_n]$) was introduced by W. Gröbner in ([Grö], 1968, 1970). Its usefulness in algebraic geometry was realized after combination with the algorithmic techniques of Gröbner bases in order to compute the dimension $\dim\mathbf{V}(I)$ of the affine algebraic set $\mathbf{V}(I)$ determined by the ideal I, or the degree of the Hilbert polynomial of the $k[x_1, ..., x_n]$-module $k[x_1, ..., x_n]/I$. The notion of (strong) independence (modulo a polynomial ideal I) was introduced in ([KW], 1989) as a *key* link between the (weak) independence (modulo I) and a Gröbner basis of I.

(ii) The reason that we call Lemma 4.4, Lemma 5.8, and Lemma 7.5 the "elimination lemma" is in the light of [Zei1] and [WZ], where an elimination lemma for the Weyl algebra was given in [Zei1] and an existence proof was due to I.N. Bernstein on the basis of holonomic module theory, and another more "effective" proof was given in [WZ] and the lemma was therefore called the "fundamental lemma" in the automatic proving of hypergeometric (ordinary and "q") multisum/integral identities (see CH.VII §2 for an introductory interpretation of the latter topic). Note that we have established the elimination lemma in a quite general extent without using any "holonomicity". As the reader will see soon in CH.VII, however, a more general holonomic module theory in connection with the function identity theory does exist, comparing with the holonomic module theory in the sense of Bernstein.

(iii) Let I be an ideal in the commutative polynomial algebra $A = k[x_1, ..., x_n]$ and $V(I)$ the affine algebraic variety defined by I. We see that the use of $\succeq_{gr}$-filtration in computing the GK-dimension of $M = A/I$ indeed recaptures the fact that $\dim V(I) = \dim V(\mathbf{LM}(I))$. And moreover, the "passing to $G^{\mathcal{F}}(A)$" trick can also be used to simplify the computation of syzygies, but we do not pursue this topic in this book.

8. Finite Dimensional Cyclic Modules

Let k be an algebraically closed field of characteristic zero, and I an ideal in the commutative polynomial algebra $k[x_1, ..., x_n]$. A well-known result in algebraic geometry states that the algebraic set $V(I)$ in the affine n-space $\mathbf{A}_k^n$ is a finite set if and only if the k-algebra $k[x_1, ..., x_n]/I$ is finite dimensional over k. Furthermore, this property of $V(I)$ has been algorithmically recognized via Gröbner bases, as recalled from (e.g., [CLO'], [BW]) below.

Theorem Let $V(I)$ be as above and fix a monomial ordering in $k[x_1, ..., x_n]$. Then the following statements are equivalent.

(i) $V(I)$ is a finite set.

(ii) For each i, $1 \le i \le n$, there is some $\alpha_i \ge 0$ such that $x_i^{\alpha_i} \in \langle \mathbf{LM}(I) \rangle$, where the latter is the ideal of $k[x_1, ..., x_n]$ generated by all leading monomials of elements in I.

(iii) Let $\mathcal{G}$ be a Gröbner basis for I. Then for each i, $1 \le i \le n$, there is some $\alpha_i \ge 0$ such that $x_i^{\alpha_i} = \mathbf{LM}(g_i)$ for some $g_i \in \mathcal{G}$.

(iv) The k-vector space $S = k\text{-Span}\{x^\beta \mid x^\beta \notin \langle \mathbf{LM}(I) \rangle\}$ is finite dimensional, where $x^\beta = x_1^{\beta_1} \cdots x_n^{\beta_n}$.

(v) The k-vector space $k[x_1, ..., x_n]/I$ is finite dimensional.

□

An immediate consequence of the above theorem is the quantitative estimate of the number of solutions of a system of equations when the number is finite.

Corollary Let $I \subset k[x_1, ..., x_n]$ be an ideal such that for each i, $1 \le i \le n$, some power $x_i^{\alpha_i} \in \langle \mathbf{LM}(I) \rangle$. Then the number of points of $V(I)$ is at most $\alpha_1 \cdot \alpha_2 \cdots \alpha_n$.

□

Concerning modules over solvable polynomial algebra with GK-dimension 0, for the sake of making the text complete, in this section we introduce the non-commutative verison of the above result obtained from [K-RW], i.e., all finite dimensional cyclic modules over an *arbitrary* solvable polynomial algebra may be recognized via left Gröbner bases. For the reader's convenience, we also include a proof here.

8.1. Theorem [K-RW] Let $A = k[a_1, ..., a_n]$ be an arbitrary solvable polynomial algebra and $(A, \mathcal{B}, \preceq)$ the left admissible system associated with A. Let $L = \langle g_1, ..., g_s]$ be a left ideal of A generated by a left Gröbner basis $\mathcal{G} = \{g_1, ..., g_s\}$. Put $M = A/L$. The following statements are equivalent.

(i) GK.dim$M = 0$;

(ii) The left A-module M is a finite dimensional k-space;

(iii) For each i, $1 \le i \le n$, there exists $g_i \in \mathcal{G}$ with that $\mathbf{LM}(g_i) = a_i^{\beta_i}$ for some $\beta_i \ge 0$.

Proof (i) ⇔ (ii) This is well-known in general.

(ii) ⇒ (iii) Note that $\mathcal{B} - \mathbf{LM}(L)$ yields a k-basis for $M = A/L$ (CH.II Theorem 3.1). For each variable a_i of A, $1 \le i \le n$, considering the sequence $\{a_i^0, a_i, a_i^2, ..., a_i^m, ...\}$ and the Gröbner representations of elements in L by $\mathcal{G}$, the assumption (ii) and the division algorithm in a solvable polynomial algebra (CH.II Proposition 7.2, Corollary 5.3) entail that there exists $g_i \in \mathcal{G}$ with

$\mathbf{LM}(g_i) = a_i^{\beta_i}$ for some $\beta_i \geq 0$.

(iii) $\Rightarrow$ (ii) Let $f \in A$ and consider the division by $\mathcal{G}$. Since $\mathcal{G}$ contains $g_1, \ldots, g_n$ with $\mathbf{LM}(g_i) = a_i^{\alpha_i}$, it is then clear that the remainder $r = \overline{f}^{\mathcal{G}}$ upon division by $\mathcal{G}$ is a linear combination of the monomials $a^\beta = a_1^{\beta_1} \cdots a_n^{\beta_n} \in \mathcal{B}$ with $0 \leq \beta_i \leq \alpha_i$, $i = 1, \ldots, n$. This shows that $M = A/L$ is a finite dimensional k-space. $\square$

8.2.   Corollary Let L be a left ideal of the solvable polynomial algebra $k[a_1, \ldots, a_n]$ such that for each i, $1 \leq i \leq n$, some power $a_i^{\alpha_i} \in \mathbf{LM}(L)$. Then $\dim_k(A/L) \leq \alpha_1 \cdot \alpha_2 \cdots \alpha_n$.

CHAPTER VI
Multiplicity Computation of Modules over Quadric Solvable Polynomial Algebras

Let $A = k[a_1, ..., a_n]$ be a quadric solvable polynomial algebra in the sense of CH.III Definition 2.1, L a left ideal of A, and $M = A/L$ the left A-module with its Hilbert polynomial

$$h_L(x) = c_d x^d + c_{d-1} x^{d-1} + \cdots + c_1 x + c_0.$$

Furthermore, let $\mathcal{G} = \{g_1, ..., g_s\}$ be a left Gröbner basis for L with leading monomials

$$\mathbf{LM}(g_j) = a_1^{\alpha_{j1}} a_2^{\alpha_{j2}} \cdots a_n^{\alpha_{jn}}, \quad j = 1, ..., s.$$

Set, as in CH.V,

$$R_j = (\alpha_{j1}, \alpha_{j2}, ..., \alpha_{jn}), \quad j = 1, ..., s.$$

The main result of this chapter is to show that

- the leading coefficient c_d of $h_L(x)$, or the multiplicity $e(M) = d! c_d$ of the module $M = A/L$, can be computed by manipulating only the data $(R_1, R_2, ..., R_n)$ given above, without computing $h_L(x)$.

Combining with CH.V, it follows that the leading term $c_d x^d$ of $h_L(x)$ can be determined completely in terms of the data $(R_1, R_2, ..., R_n)$.

1. The Multiplicity $e(M)$ of a Module M

Let $A = k[a_1, ..., a_n]$ be a quadric solvable polynomial k-algebra as before. If L is a proper left ideal of A and $M = A/L$, then it follows from CH.V §7 that

$$\text{GK.dim}(M) = \deg h_L(x),$$

where $h_L(x)$ is the Hilbert polynomial of the A-module M. Moreover, write

$$h_L(x) = c_d x^d + c_{d-1} x^{d-1} + \cdots + c_1 x + c_0, \ c_i \in \mathbb{Q}.$$

Then since L is a proper left ideal, c_d, the leading coefficient of $h_L(x)$, is a positive rational number (CH.V Theorem 7.4). We call $d!c_d$ the *multiplicity* of the left A-module M and denote it by $e(M)$. The aim of this section is to interpret the name of $e(M)$ from a viewpoint of representation theory of A. (The reader may compare $e(M)$ defined here with the multiplicity of a module over a commutative Noetherian local ring in commutative algebra (algebraic geometry) by refering to, e.g., [Eis] ch.12.) Note that for any finitely generated A-module $N = \sum_{i=1}^{s} A\xi_i$ with $\xi_i \in N$, $\text{GK.dim} N = \max\{\text{GK.dim}(A\xi_i) \mid i = 1, ..., s\}$ (cf. [KL]). The interpretation given in this section is valid for all finitely generated A-modules.

The first obvious fact is that if $\text{GK.dim} M = 0$, then M is a finite dimensional k-space of $\dim_k M = c_d = e(M)$, and hence the length of any composition series of M is $\le e(M)$.

If $\text{GK.dim} M = d > 0$, then M is infinite dimensional over k. Suppose that M is *d-pure* in the sense that for every sequence of nonzero A-submodules $V \subset W \subset M$, the euqality

$$\text{GK.dim}(W/V) = \text{GK.dim} W = d$$

holds. If we consider the filtrations on W and W/V induced by the filtration FM, which is induced by the standard filtration FA on A, the following result is easily proved.

1.1. Proposition If $M = A/L$ is d-pure, then M has finite length and any composition series of M has length $\le e(M)$. Therefore, if $e(M) = 1$, then M is irreducible.

$\square$

Let S be an A-module. Recall from homological algebra that the number

$$j_A(S) = \min\left\{i \in \mathbb{N} \mid \text{Ext}_A^i(S, A) \ne 0\right\}$$

is called the *grade number* of S. (For a right module, there is a similar definition.) A is said to be an *Auslander regular algebra* if A is left and right Noetherian with finite global homological dimension gl.dim$A = \omega$, and for each nonzero finitely generated A-module M, each $i \geq 0$ and each nonzero submodule $N \subset \text{Ext}^i_A(A, M)$, the inequality $j_A(N) > i$ holds. If furthermore the equality

$$(*) \qquad\qquad j_A(M) + \text{GK.dim}M = n = \text{GK.dim}A$$

holds for every nonzero finitely generated module M, then A is said to have the *Cohen-Macaulay property*. If $j_A(M) = \omega - \text{gl.dim}A$, then M is called a *holonomic* A-module (see CH.VII §2 for a background introduction to holonomic modules). Any holonomic module over an Auslander regular algebra is of finite length (e.g., see [Li1], [LVO4]). If the above $(*)$ holds for A, then any holonomic A-module M is d-pure where $d = n - \omega$.

Example Let A be one of the following k-algebras or their associated Graded and Rees algebras with respect to the standard filtration:
(i) The nth Weyl algebra $A_n(k)$;
(ii) The enveloping algebra $U(\mathbf{g})$ of a finite dimensional Lie algebra $\mathbf{g}$;
(iii) The additive analogue $A_n(q_1, ..., q_n)$ of the Weyl algebra;
(iv) Any homogeneous solvable polynomial algebra;
(v) The q-Heisenberg algebra $\mathbf{h}_n(q)$.
Then A is a quadric solvable polynomial algebra satisfying the above $(*)$. (We refer to CH.I §5, [Lev], [Li1], [LVO1], [LVO2] and [LVO4] for the reason why $(*)$ holds for these algebras.)

2. Computation of $e(M)$

Let $A = k[a_1, ..., a_n]$ be an arbitrary quadric solvable polynomial k-algebra with the associated (left) admissible system $(A, \mathcal{B}, \succeq_{gr})$, where

$$\mathcal{B} = \left\{ a^\alpha = a_1^{\alpha_1} \cdots a_n^{\alpha_n} \ \Big| \ \alpha = (\alpha_1, ..., \alpha_n) \in \mathbb{N}^n \right\}$$

and $\succeq_{gr}$ is some graded monomial ordering on $\mathcal{B}$, and let L be a proper left ideal of A with a fixed Gröbner basis $\mathcal{G} = \{g_1, ..., g_s\}$. Suppose that the Hilbert polynomial of the left A-module $M = A/L$ is given by

$$h_L(x) = c_d x^d + c_{d-1} x^{d-1} + \cdots + c_1 x + c_0.$$

This section is devoted to the computation of the multiplicity $e(M) = d!c_d$ for M, without computing $h_L(x)$.

We first deal with *homogeneous* solvable polynomial algebras, i.e., we assume that A is a homogeneous solvable polynomial algebra. All notation are retained from previous chapters.

Since L is a proper left ideal of A and $\mathcal{G} = \{g_1, ..., g_s\}$ is a Gröbner basis of L, then it follows from CH.V §2 that A/L, $A/\langle \mathbf{LM}(L)]$, and $A/\langle \mathbf{LM}(\mathcal{G})]$ have the same Hilbert polynomial $h_L(x)$, where $\langle \mathbf{LM}(L)]$ is the monomial left ideal of A generated by the leading monomials of all elements in L and $\langle \mathbf{LM}(\mathcal{G})]$ is the monomial left ideal of A generated by $\mathbf{LM}(\mathcal{G}) = \{\mathbf{LM}(g_1), ..., \mathbf{LM}(g_s)\}$ (note that $\langle \mathbf{LT}(L)] = \langle \mathbf{LT}(\mathcal{G})]$ by CH.V §1). For the left A-module $M = A/L$, in order to compute $e(M) = d!c_d$ without computing $h_L(x)$, it is therefore sufficient to assume that L *is a monomial* left ideal, say $L = \langle m_1, ..., m_s]$ where $m_1, ..., m_s \in \mathcal{B}$. Thus $\mathcal{G} = \{m_1, ..., m_s\}$ is a left Gröbner basis for L by CH.V §1.

With the fixed notation and assumption as above, the computation of $e(M)$ will stem from another constructive proof of the existence of the Hilbert polynomial $h_L(x)$ of M by using $\mathcal{G}$. (Compare with CH.V §3, and see [CLO'] CH.9 for the constructive proof of the existence of the Hilbert polynomial for a polynomial ideal in the commutative case, which is based on Hilbert's famous article *Über die Theorie der Algebraischen Formen* (1890)).
Adopting similar notation as in [CLO'] CH.9, we start by writing

$$\mathbf{C}(L) = \left\{ \alpha = (\alpha_1, ..., \alpha_n) \in I\!N^n \ \Big| \ a^\alpha \in \mathcal{B}, \ a^\alpha \notin L \right\}$$

for the complement of L in $\mathcal{B}$, i.e., $\mathbf{C}(L) = \mathcal{B} - L$, and writing

$$\begin{aligned}
e_1 &= (1, 0, ..., 0) \\
e_2 &= (0, 1, ..., 0) \\
&\ \ \vdots \\
e_n &= (0, 0, ..., 1)
\end{aligned}$$

where each e_i is called a *unit vector.*

2.1. Definition (i) For $\{e_{i_1}, ..., e_{i_r}\} \subset \{e_1, ..., e_n\}$ with $i_1 < \cdots < i_r$, $r \leq n$, the subset

$$[e_{i_1}, ..., e_{i_r}] = \left\{ \sum_{j=1}^{r} q_j e_{i_j} \ \Big| \ q_j \in I\!N \text{ for } 1 \leq j \leq r \right\} \subset I\!N^n$$

is called an r-dimensional *coordinate subspace* of $I\!N^n$ determined by $e_{i_1}, ..., e_{i_r}$.
(ii) For $\beta = \sum_{j \notin \{i_1, ..., i_r\}} q_j e_j \in I\!N^n$ with $q_j \in I\!N$, the subset

$$\beta + [e_{i_1}, ..., e_{i_r}] = \left\{ \beta + \gamma \ \Big| \ \gamma \in [e_{i_1}, ..., e_{i_r}] \right\}$$

is called a *translate* of the r-dimensional coordinate subspace $[e_{i_1}, ..., e_{i_r}]$.

2.2. Lemma Let $\beta + [e_{i_1}, ..., e_{i_r}]$ be a translate of the coordinate subspace $[e_{i_1}, ..., e_{i_r}] \subset I\!N^n$ with $\beta = \sum_{j \notin \{i_1,...,i_r\}} q_j e_j$.
(i) If $s > |\beta|$, the number of points $\beta + \gamma$ in $\beta + [e_{i_1}, ..., e_{i_r}]$ with $|\beta| + |\gamma| \leq s$ is equal to

$$\binom{r + s - |\beta|}{s - |\beta|,}.$$

(ii) For $s > |\beta|$, the number of points obtained in (i) above is a polynomial function of s of degree r, and the coefficient of s^r is $\dfrac{1}{r!}$.

$\square$

The next proposition is a noncommutative analogue of ([CLO'] Ch.9 Theorem 3). Though its proof is similar to that given in the commutative case, to see how the Noetherian property is essential in the proof (as that for Buchberger's Algorithm) and why a similar argumentation works for a noncommutative homogeneous solvable polynomial algebra, it is worthwhile to write down the detailed proof here.

2.3. Proposition With notation as above, if L is a proper monomial left ideal, then the set $\mathbf{C}(L) \subset I\!N^n$ can be written as a disjoint union

$$\mathbf{C}(L) = C_0 \cup C_1 \cup \cdots \cup C_{n-1} \cup C_n,$$

where each C_i is a finite (not necessarily disjoint) union of translates of i-dimensional coordinate subspaces of $I\!N^n$. And $C_n \neq \emptyset$ if and only if $L = \{0\}$.

Proof Suppose $L \neq \{0\}$. We do induction on the number of variables n. If $n = 1$, then $A = k[a_1]$ is the polynomial k-algebra in one variable and $L = \langle a_1^\ell \rangle$ for some integer $\ell > 0$. The only monomials not contained in L are $1, a_1, ..., a_1^{\ell-1}$. Hence, $\mathbf{C}(L) = \{0\} \cup \{1\} \cup \cdots \cup \{\ell - 1\}$, where each $\{j\}$ is a 0-dimensional coordinate subspace of $I\!N$ (a translate of the origin $\{0\}$).
Assume that the result holds for homogeneous solvable polynomial algebras of $n - 1$ generators. Then the result holds for the subalgebra $k[a_1, ..., a_{n-1}]$ of A. Let L be a monomial left ideal of A. For each $j \geq 0$, let L_j be the left ideal in $k[a_1, ..., a_{n-1}]$ generated by monomials $a^\alpha = a_1^{\alpha_1} \cdots a_{n-1}^{\alpha_{n-1}}$ with the property that $a^\alpha \cdot a_n^j \in L$. Then it follows from CH.V §1 that

$$\mathbf{C}(L_j) = \left\{ \beta \in I\!N^{n-1} \,\middle|\, a^\beta \cdot a_n^j \notin L \right\}.$$

If $j < j' = j + s$ and if $a^\alpha \in L_j$, then since $a^\alpha \cdot (a_n^s a_n^j) = \lambda_{\alpha,s} a_n^s a^\alpha \cdot a_n^j \in L$ for some $\lambda_{\alpha,s} \in k - \{0\}$, we have $L_j \subset L_{j'}$. Thus, there exists an integer j_0 such that $L_j = L_{j_0}$ for all $j \geq j_0$ (note that solvable polynomial algebras are Noetherian domains). By the induction hypothesis, for each of the L_j, $0 \leq j \leq j_0$, there is a decomposition

$$\mathbf{C}(L_j) = C_0^j \cup C_1^j \cup \cdots \cup C_{n-1}^j,$$

where C_k^j is a finite (not necessarily disjoint) union of translates of k-dimensional coordinate subspaces in $I\!N^{n-1}$.

For each $j \in I\!N$, write

$$C_k^j \times \{j\} = \left\{ (\beta, j) \in I\!N^n \mid \beta \in C_k^j \subset I\!N^{n-1} \right\}.$$

We now claim that

$(*)$ $\qquad\qquad\qquad$ $\mathbf{C}(L) = C_0 \cup C_1 \cup \cdots \cup C_{n-1} \cup C_n$,

where

$$C_{k+1} = \left(C_k^{j_0} \times I\!N \right) \cup \bigcup_{j=0}^{j_0-1} \left(C_{k+1}^j \times \{j\} \right)$$

for $0 \le k \le n-1$, and

$$C_0 = \mathbf{C}(L) - \left(\bigcup_{k=1}^{n} C_k \right).$$

To see this, first note that $C_k^j \times \{j\} \subset \mathbf{C}(L)$ by the definition of $\mathbf{C}(L_j)$. To show that $C_k^{j_0} \times I\!N \subset \mathbf{C}(L)$, note that $L_j = L_{j_0}$ for $j \ge j_0$. Hence it is clear that $C_k^{j_0} \times \{j\} \subset \mathbf{C}(L)$ for $j \ge j_0$'s. For $j < j_0$, we have $a^\beta \cdot a_n^j \notin L$ whenever $a^\beta \cdot a_n^{j_0} \notin L$ since A is a homogeneous solvable polynomial algebra and L is a left ideal of A. This shows that $C_k^{j_0} \times \{j\} \subset \mathbf{C}(L)$ for $j < j_0$. Thus, $C_{k+1} \subset \mathbf{C}(L)$ for all $k \ge 0$, and it follows from the definition of C_0 that the above $(*)$ holds. Finally, we show that each C_k is a finite union of translates of k-dimensional coordinate subspaces in $I\!N^n$. But this is clear for $k > 0$. For $k = 0$, suppose that there were a point $\alpha = (\alpha_1, ..., \alpha_n) \in C_0$ with $\alpha_n \ge j_0$. Then $\alpha \in \mathbf{C}(L_{j_0}) \times \{\alpha_n\}$ and consequently $\alpha \in C_k^{j_0} \times I\!N \subset C_{k+1}$ for some k. This contradicts the definition of C_0. Hence we must have $\alpha_n < j_0$ for all points of C_0. It turns out that if C_0 were infinite, there would be some $j < j_0$ such that infinitely many points of C_0 would lie in $\mathbf{C}(L_j) \times \{j\}$ because A is a homogeneous solvable polynomial algebra. Note that C_0^j is finite by our inductive hypothesis. It follows that some of these points would have to be in $C_k^j \times \{j\}$ for some $k > 0$. But the latter set is contained in C_k for $k > 0$ and this again contradicts the definition of C_0. This shows that C_0 must be finite, finishing the proof. $\qquad$ $\square$

To go further, let us write the members of the given left Gröbner basis $\mathcal{G} = \{m_1, ..., m_s\}$ of L as

(1)
$$\begin{cases} m_1 & = & a_1^{\alpha_{11}} a_2^{\alpha_{12}} \cdots a_n^{\alpha_{1n}}, \\ m_2 & = & a_1^{\alpha_{21}} a_2^{\alpha_{22}} \cdots a_n^{\alpha_{2n}}, \\ \vdots & & \vdots \\ m_s & = & a_1^{\alpha_{s1}} a_2^{\alpha_{s2}} \cdots a_n^{\alpha_{sn}}. \end{cases}$$

Recall from CH.V Theorem 4.2 that if we put

$$R_k = (\alpha_{k1}, ..., \alpha_{kn}), \quad k = 1, ..., s,$$

$$M_k = \left\{ i \in \{1, ..., n\} \mid \alpha_{ki} \neq 0 \text{ in } R_k \right\}, \quad k = 1, ..., s,$$

$$\mathcal{M} = \left\{ J \subset \{1, ..., n\} \mid J \cap M_k \neq \emptyset, \ k = 1, ..., s \right\},$$

then $\mathrm{GK.dim}\, M = n - \min\{|J| \mid J \in \mathcal{M}\}$, where $M = A/L$.

Since A is a homogeneous solvable polynomial algebra, by the definition of $\mathbf{C}(L)$, CH.V §1, and Proposition 2.3, an easy but very useful observation is recorded as follows.

2.4. Observation With notation as before, let $\beta + [e_{i_1}, ..., e_{i_r}]$ be a translate contained in C_r with $\beta = \sum_{j \notin \{i_1, ..., i_r\}} q_j e_j$. Putting $J = \{1, ..., n\} - \{i_1, ..., i_r\}$, then $J \in \mathcal{M}$.

$\square$

2.5. Theorem With notation as above, for $M = A/L$, if $\mathrm{GK.dim}\, M = d$, then the following hold.
(i) $\mathbf{C}(L) = C_0 \cup C_1 \cup \cdots \cup C_d$, with $C_d \neq \emptyset$, where each C_i is a finite (not necessarily disjoint) union of translates T_{i_j} of i-dimensional coordinate subspaces in $I\!N^n$, i.e.,

(2). $$C_i = T_{i_1} \cup T_{i_2} \cup \cdots \cup T_{i_m}.$$

(ii) For each $s \geq 0$ and each C_i, T_{i_j} in the above (2), if we put

$$C_i^s = \left\{ a^\alpha \in C_i \mid |\alpha| \leq s \right\},$$

$$T_{i_j}^s = \left\{ a^\alpha \in T_{i_j} \mid |\alpha| \leq s \right\},$$

then $|C_i^s|$ is a polynomial of degree i in s, in particular, $|C_d^s|$ is a polynomial of degree d when s is large enough, and the leading term of the Hilbert polynomial $h_L(x)$ is given by the leading term of the polynomial $|C_d^s|$ which is of the form $\frac{N}{d!} t^d$, where N is the number of *distinct* T_{d_j} appearing in the above decomposition (2).

Proof (i) If $\mathrm{GK.dim}\, M = d$, then there is some $J \in \mathcal{M}$ with $|J| = n - d$ such that $J \cap M_k \neq \emptyset$, $k = 1, ..., s$. Put $\{i_1, ..., i_d\} = \{1, ..., n\} - J$. By CH.V §1, it is clear that $[e_{i_1}, ..., e_{i_d}] \subset C_d$, i.e., $C_d \neq \emptyset$. If $d < n$, then note that since $n - d$ is the smallest number of $|J|$ with $J \in \mathcal{M}$, Observation 2.4 yields $C_{d+i} = \emptyset$ for $i > 0$.

(ii) First note that for $s \gg 0$, $\dim_k(F_s A + L/L) = |\mathbf{C}(L)_{\leq s}| = h_L(s)$, where $h_L(x)$ is the Hilbert polynomial of M, FA is the standard filtration of A, and $\mathbf{C}(L)_{\leq s} = \mathbf{C}(L) \cap F_s A$. When Lemma 2.2 applies to the decomposition (2) in part (i), the counting principle (inclusion and exclusion for finite sets from elementary combinatorics) yields the desired result. $\square$

From the last theorem it is then clear that $e(M) = N$. We now proceed to compute the number N without computing $h_L(x)$.

For *each* $J \in \mathcal{M}$ with $J = \{\ell_1, ..., \ell_{n-d}\}$ where $d = \mathrm{GK.dim}M$, put $\{i_1, ..., i_d\} = \{1, ..., n\} - J$. Assume that $\ell_1 < \ell_2 < \cdots < \ell_{n-d}$ and $i_1 < i_2 < \cdots < i_d$. Since A is homogeneous solvable, by CH.V §1, every m_k in the foregoing (1) may be written as

$$(1') \quad m_k = \mu \cdot \left(a_{\ell_1}^{\alpha_{k\ell_1}} a_{\ell_2}^{\alpha_{k\ell_2}} \cdots a_{\ell_{n-d}}^{\alpha_{k\ell_{n-d}}} \right) \cdot \left(a_{i_1}^{\alpha_{ki_1}} a_{i_2}^{\alpha_{ki_2}} \cdots a_{i_d}^{\alpha_{ki_d}} \right), \quad k = 1, ..., s,$$

for some $\mu \in k - \{0\}$. To be convenient, let us use the above (1') to associate $\mathcal{G}$ with an $s \times n$ matrix

$$M(\mathcal{G}, J) = \begin{pmatrix} \alpha_{1\ell_1} & \cdots & \alpha_{1\ell_{n-d}} & \alpha_{1i_1} & \cdots & \alpha_{1i_d} \\ \vdots & & \vdots & \vdots & & \vdots \\ \alpha_{s\ell_1} & \cdots & \alpha_{s\ell_{n-d}} & \alpha_{si_1} & \cdots & \alpha_{si_d} \end{pmatrix}$$

and call $M(\mathcal{G}, J)$ the *matrix of $\mathcal{G}$ determined by J*.
For each $p = 1, 2, ..., n - d$, put

$$L_p = J - \{\ell_p\}, \quad p = 1, ..., n - d.$$

Since $\mathrm{GK.dim}M = d$ and $J \cap M_j \neq \emptyset$, $j = 1, ..., s$, there exists some M_k such that $L_p \cap M_k = \emptyset$, and consequently, the kth row of $M(\mathcal{G}, J)$, denoted R_k, has the form

$$R_k = (0, ..., 0, \alpha_{k\ell_p}, 0, ..., 0, \gamma_k)$$

with $\alpha_{k\ell_p} \neq 0$ and $\gamma_k = (\alpha_{ki_1}, ..., \alpha_{ki_d}) \in [e_{i_1}, ..., e_{i_d}]$. Putting

$$(**) \qquad \alpha_{\ell_p} = \min\left\{ \alpha_{k\ell_p} \mid L_p \cap M_k = \emptyset \right\}, \quad p = 1, ..., n - d,$$

and interchanging the rows of $M(\mathcal{G}, J)$ (if necessary), the matrix $M(\mathcal{G}, J)$ now has the configuration

$$CM(\mathcal{G}, J) = \begin{pmatrix} \alpha_{\ell_1} & 0 & \cdots & 0 & \gamma_1 \\ 0 & \alpha_{\ell_2} & \cdots & 0 & \gamma_2 \\ \vdots & \vdots & \vdots & \vdots & \vdots \\ 0 & 0 & \cdots & \alpha_{\ell_{n-d}} & \gamma_{n-d} \\ \alpha_{n-d+1,\ell_1} & \alpha_{n-d+1,\ell_2} & \cdots & \alpha_{n-d+1,\ell_{n-d}} & \gamma_{n-d+1} \\ \vdots & \vdots & \vdots & \vdots & \vdots \\ \alpha_{s\ell_1} & \alpha_{s\ell_2} & \cdots & \alpha_{s\ell_{n-d}} & \gamma_s \end{pmatrix}$$

2.6. Proposition With notation as above, the following holds.

(i) Let $\{\alpha_{\ell_p} \mid p = 1, ..., n - d\}$ be defined by some $J \in \mathcal{M}$ as in $(**)$ above. If $\beta + [e_{i_1}, ..., e_{i_d}]$ is a translate of some d-dimensional coordinate subspace $[e_{i_1}, ..., e_{i_d}]$ contained in C_d with $\beta = \sum_{p \notin \{i_1, ..., i_d\}} q_p e_p$, then $q_p < \alpha_{\ell_p}$. It follows that, if we write E for the number of *distinct* $J \in \mathcal{M}$ with $|J| = n - d$, then

$$e(M) \leq \sum_{p=1}^{E} \prod^{n-d} \alpha_{\ell_p},$$

where each product $\prod_{p=1}^{n-d} \alpha_{\ell_p}$ is determined by some $J \in \mathcal{M}$ with $|J| = n - d$.

(ii) $e(M)$ can be computed by using only the entries of the associated $s \times n$ matrix $M(\mathcal{G}, J)$ of $\mathcal{G}$ determined by *some chosen* $J \in \mathcal{M}$ with $|J| = n - d$, without computing the Hilbert polynomial $h_L(x)$ of M. More precisely, let $\{\alpha_{\ell_p} \mid p = 1, ..., n - d\}$ be defined by J as in $(**)$ above. Considering the last $s - n + d$ rows of $CM(\mathcal{G}, J)$ as resulted above, if we set, for $i = 1, ..., s - n + d$,

$$\alpha(n - d + i) = \left(\alpha_{n-d+i, \ell_1}, \alpha_{n-d+i, \ell_2}, ..., \alpha_{n-d+i, \ell_{n-d}}, 0, ..., 0 \right),$$

then

$$e(M) = \left| \left\{ \beta = \sum_{p=1}^{n} {}^{d} q_{\ell_p} e_{\ell_p} \;\middle|\; \begin{array}{l} q_{\ell_p} < \alpha_{\ell_p}, \; p = 1, ..., n - d, \\ \alpha(n - d + i) \not\succeq' \beta, \; i = 1, ..., s - n + d \end{array} \right\} \right|,$$

where $\succeq'$ is the Dickson partial order on $\mathbb{N}^n$ induced by $\succeq_{gr}$ or by the divisibility in A (see CH.II §6 and the proof of CH.II Proposition 7.2).

Proof (i) Since $\beta + [e_{i_1}, ..., e_{i_d}]$ is contained in C_d, the first part of the statement follows from CH.V §1 by looking at the configuration $CM(\mathcal{G}, J)$ of $M(\mathcal{G}, J)$ obtained above, and the inequality follows from Observation 2.4. and the rule of product.

(ii) For the chosen $J \in \mathcal{M}$ with $|J| = n - d$ and $J = \{\ell_1, ..., \ell_{n-d}\} = \{1, ..., n\} - \{i_1, ..., i_d\}$, consider $\beta = \sum_{\ell_p \in J} q_{\ell_p} e_{\ell_p}$ with $q_{\ell_p} < \alpha_{\ell_p}$. By the foregoing discussion and CH.V §1, we easily see that $e(M)$ is nothing but the number of all a^β's which cannot be divided by any one of the monomials:

$$m'_{n-d+1} = a_{\ell_1}^{\alpha_{n-d+1, \ell_1}} a_{\ell_2}^{\alpha_{n-d+1, \ell_2}} \cdots a_{\ell_{n-d}}^{\alpha_{n-d+1, \ell_{n-d}}}$$

$$m'_{n-d+2} = a_{\ell_1}^{\alpha_{n-d+2, \ell_1}} a_{\ell_2}^{\alpha_{n-d+2, \ell_2}} \cdots a_{\ell_{n-d}}^{\alpha_{n-d+2, \ell_{n-d}}}$$

$$\vdots \qquad \vdots \qquad \vdots$$

$$m'_s = a_{\ell_1}^{\alpha_{s\ell_1}} a_{\ell_2}^{\alpha_{s\ell_2}} \cdots a_{\ell_{n-d}}^{\alpha_{s\ell_{n-d}}}$$

□

Furthermore, let $J \in \mathcal{M}$ with $J = \{\ell_1, ..., \ell_{n-d}\} = \{1, ..., n\} - \{i_1, ..., i_d\}$. For $p = 1, ..., n - d$, put

$$(* * *) \quad \alpha'_{\ell_p} = \min \left\{ \alpha_{k\ell_p} \mid \alpha_{k\ell_p} \neq 0 \text{ in the above } CM(\mathcal{G}, J), \ k = 1, ..., s \right\}.$$

Again by an easy combinitorial computation and combining Proposition 2.6, we can mention the following estimation formula for multiplicity.

2.7. Proposition With notation as above, we have

$$\sum^{E} \prod_{p=1}^{n-d} \alpha_{\ell_p} \geq e(M) \geq \sum^{E} \prod_{p=1}^{n-d} \alpha'_{\ell_p},$$

where each product $\prod_{p=1}^{n-d} \alpha_{\ell_p}$, respectively $\prod_{p=1}^{n-d} \alpha'_{\ell_p}$, is determined by some $J \in \mathcal{M}$ with $|J| = n - d$, and each α_{ℓ_p}, respectively each α'_{ℓ_p}, is as in the foregoing $(**)$, respectively in the foregoing $(***)$. The equalities hold in the above inequalities if $\alpha_{\ell_p} = \alpha'_{\ell_p}$, in particular, if the given Gröbner basis $\mathcal{G}$ of L consists of $s = n - d$ monomials.

$\square$

The results established above have two immediate consequences in certain special cases.

2.8.   Corollary Let L, $\mathcal{G}$, and $M = A/L$ be as before. With notation as above, $e(M) = 1$ if and only if there is only one $J \in \mathcal{M}$ with $|J| = n - d$, say $J = \{\ell_1, ..., \ell_{n-d}\} \subset \{1, ..., n\}$, such that $\alpha_{\ell_p} = 1$, where α_{ℓ_p} is as in the foregoing $(**)$, $p = 1, ..., n - d$.

$\square$

2.9.   Corollary Let L, $\mathcal{G}$, and $M = A/L$ be as before. Suppose GK.dim$M = n - 1$. If $C_i = \{\alpha_{1i}, ..., \alpha_{si}\}$ denotes the ith column of $M(\mathcal{G}, J)$ for some fixed $J \in \mathcal{M}$ with $|J| = 1$, then $e(M) = \sum \alpha_i$, where

$$\alpha_i = \min \{\alpha_{1i}, ..., \alpha_{si}\}, \quad i = 1, ..., n.$$

$\square$

Summing up, let K be an *arbitrary* left ideal of the homogeneous solvable polynomial algebra A, $S = A/K$. If $\mathcal{G} = \{g_1, ..., g_s\}$ is a left Gröbner basis of K, then we write $\langle \mathbf{LM}(\mathcal{G})]$ for the monomial left ideal generated by the leading monomials $m_1 = \mathbf{LM}(g_1), ..., m_s = \mathbf{LM}(g_s)$. Put $L = \langle \mathbf{LM}(\mathcal{G})]$ in Proposition 2.6. Then it follows from previous 2.6–2.7 that the following theorem holds.

2.10. Theorem Let K and S be as above, and suppose GK.dim$S = d$. With notation as before, the following hold.

(i)

$$\sum_{p=1}^{E} \prod_{p=1}^{n-d} \alpha_{\ell_p} \geq e(S) \geq \sum_{p=1}^{E} \prod_{p=1}^{n-d} \alpha'_{\ell_p},$$

where each product $\displaystyle\prod_{p=1}^{n-d} \alpha_{\ell_p}$, respectively $\displaystyle\prod_{p=1}^{n-d} \alpha'_{\ell_p}$, is determined by some $J \in \mathcal{M}$ with $|J| = n - d$, and each α_{ℓ_p}, respectively each α'_{ℓ_p}, is as in the foregoing (**), respectively as in the foregoing (***). The equalities hold in the above inequalities if $\alpha_{\ell_p} = \alpha'_{\ell_p}$, in particular, if the given left Gröbner basis $\mathcal{G}$ of K consists of $n - d$ elements.

(ii) $e(S)$ can be computed by manipulating only the entries of the associated matrix $M(\mathcal{G}, J)$ of $\mathcal{G}$ determined by some $J \in \mathcal{M}$ with $|J| = n - d$, as in the proof of Proposition 2.6(ii).

(iii) $e(S) = 1$ if and only if there is only one $J \in \mathcal{M}$ with $J = \{\ell_1, ..., \ell_{n-d}\}$ such that $\alpha_{\ell_p} = 1$, where α_{ℓ_p} is as in the foregoing (**), $p = 1, ..., n - d$.

(iv) Suppose GK.dim$S = n - 1$. If $C_i = \{\alpha_{1i}, ..., \alpha_{si}\}$ denotes the ith column of $M(\mathcal{G}, J)$ for some fixed $J \in \mathcal{M}$ with $|J| = 1$, then $e(S) = \sum \alpha_i$, where for $i = 1, ..., n$

$$\alpha_i = \min\{\alpha_{1i}, ..., \alpha_{si}\}, \quad i = 1, ..., n.$$

$\square$

Now we return to an *arbitrary* quadric solvable polynomial algebra $A = k[a_1, ..., a_n]$. As dealing with GK-dimension of modules in CH.V §7, we establish the computational result of $e(M)$ for the module $M = A/L$, where L is a proper left ideal of A, by passing to the homogeneous solvable polynomial algebra $G^{\mathcal{F}}(A)$ with respect to the $\succeq_{gr}$-filtration $\mathcal{F}A$ on A.

Fix a left Gröbner basis $\mathcal{G} = \{g_1, ..., g_s\}$ for L. Consider the filtrations on L and $M = A/L$ induced by the filtration $\mathcal{F}A$ on A as in CH.V §7. It follows from CH.V Proposition 6.1, Lemma 7.1, and Proposition 7.2 that $G^{\mathcal{F}}(A)$ is a homogeneous solvable polynomial algebra and that $\sigma(\mathcal{G}) = \{\sigma(\mathbf{LM}(g_1)), ..., \sigma(\mathbf{LM}(g_s))\}$ is a left Gröbner basis for $G^{\mathcal{F}}(L)$ consisting of monomials, say

$$\sigma(\mathbf{LM}(g_i)) = \mathbf{LM}(\sigma(g_i)) = \sigma(a_1)^{\alpha_{i1}} \sigma(a_2)^{\alpha_{i2}} \cdots \sigma(a_n)^{\alpha_{in}}, \quad i = 1, ..., s.$$

Moreover, the argumentation of CH.V §7 entails that M and $G^{\mathcal{F}}(M) = G^{\mathcal{F}}(A)/G^{\mathcal{F}}(L)$ have the same Hilbert polynomial $h_L(x)$. Thus, putting in Theorem 2.10

$$\begin{aligned} K &= G^{\mathcal{F}}(L), \quad S = G^{\mathcal{F}}(A)/G^{\mathcal{F}}(L), \\ m_1 &= \mathbf{LM}(\sigma(g_1)), m_2 = \mathbf{LM}(\sigma(g_2)), ..., m_s = \mathbf{LM}(\sigma(g_s)), \end{aligned}$$

we are now able to state the main result of this section, as follows.

2.11. Theorem With notation as before, suppose GK.dim$M = d$. Then the following holds.

(i) $e(M)$ can be computed by manipulating only the entries of the associated matrix $M(\mathcal{G}, J)$ of $\mathcal{G}$ determined by some $J \in \mathcal{M}$ with $|J| = n - d$, as in the proof of Proposition 2.6(ii).

(ii) $e(M) = 1$ if and only if there is only one $J \in \mathcal{M}$ with $J = \{\ell_1, ..., \ell_{n-d}\}$ such that $\alpha_{\ell_p} = 1$, where α_{ℓ_p} is as in the foregoing $(**)$, $p = 1, ..., n - d$.

(iii) Suppose GK.dim$S = n - 1$. If $C_i = \{\alpha_{1i}, ..., \alpha_{si}\}$ denotes the ith column of $M(\mathcal{G}, J)$ for some fixed $J \in \mathcal{M}$ with $|J| = 1$, then $e(M) = \sum \alpha_i$, where for $i = 1, ..., n$

$$\alpha_i = \min\{\alpha_{1i}, ..., \alpha_{si}\}, \quad i = 1, ..., n.$$

3. Computation of GK.dim$(M \otimes_k N)$ and $e(M \otimes_k N)$

Let $A = k[a_1, ..., a_n]$ and $B = k[b_1, ..., b_m]$ be two quadric solvable polynomial k-algebras with the associated (left) admissible systems $(A, \mathcal{B}(A), \succeq_{gr})$ and $(B, \mathcal{B}(B), \succeq_{gr})$, respectively. In this section we compute GK.dim$(M \otimes_k N)$ and $e(M \otimes_k N)$ for the left $A \otimes_k B$-module $M \otimes_k N$, where $M = A/L$ for some left ideal L of A and $N = B/K$ for some left ideal K of B, and $(a \otimes b)(x \otimes y) = (ax \otimes by)$ for $a \otimes b \in A \otimes_k B$, $x \otimes y \in M \otimes_k N$.

First note that since k is a field, under the canonical k-algebra homomorphisms $\varphi_A \colon A \to A \otimes_k B$, $\varphi_A(a) = a \otimes 1$, and $\varphi_B \colon B \to A \otimes_k B$, $\varphi_B(b) = 1 \otimes b$, we have $A \cong \varphi_A(A)$ and $B \cong \varphi_B(B)$ (e.g., see [Pie] CH.9). Hence, A and B may be viewed as subalgebras of $A \otimes_k B$, and each $a \in A$, respectively, each $b \in B$, may be identified with $a \otimes 1$, respectively, with $1 \otimes b$. Moreover, there is the left $A \otimes_k B$-module isomorphism

$$M \otimes_k N = \frac{A}{L} \otimes_k \frac{B}{K} \cong \frac{A \otimes_k B}{\langle L, K]},$$

where $\langle L, K]$ is the left ideal of $A \otimes_k B$ generated by L and K, identified with $\varphi_A(L)$ and $\varphi_B(K)$ respectively.

Secondly, since A and B are quadric solvable algebras, i.e.,

$$a_j a_i = \lambda_{ji} a_i a_j + \sum \lambda_{ji}^{k\ell} a_k a_\ell + \sum \lambda_h a_h + \lambda, \ 1 \le i < j \le n,$$
$$\text{where } \lambda_{ji} \ne 0, \ \lambda_{ji}^{k\ell}, \ \lambda_h, \ \lambda \in k;$$
$$b_p b_q = \mu_{pq} b_q b_p + \sum \mu_{pq}^{uv} a_u a_v + \sum \mu_t a_t + \mu, \ 1 \le q < p \le m,$$
$$\text{where } \mu_{pq} \ne 0, \ \mu_{pq}^{uv}, \ \mu_t, \ \mu \in k,$$

putting $T = A \otimes_k B$ and

$$X_i = a_i \otimes 1, \ i = 1, ..., n, \quad Y_j = 1 \otimes b_j, \ j = 1, ..., m,$$

then $T = k[X_1, ..., X_n, Y_1, ..., Y_m]$ satisfies

$$X_j X_i = \lambda_{ji} X_i X_j + \sum \lambda_{ji}^{k\ell} X_k X_\ell + \sum \lambda_h X_h + \lambda, \ 1 \leq i < j \leq n,$$
$$Y_p Y_q = \mu_{pq} Y_q Y_p + \sum \mu_{pq}^{uv} Y_u Y_v + \sum \mu_t Y_t + \mu, \ m \geq p > q \geq 1,$$
$$Y_j X_i = X_i Y_j, \ 1 \leq i \leq n, \ 1 \leq j \leq m,$$

and the set of standard monomials

$$\mathcal{B}(T) = \left\{ X^\alpha Y^\beta = X_1^{\alpha_1} \cdots X_n^{\alpha_n} Y_1^{\beta_1} \cdots Y_m^{\beta_m} \ \middle| \ \begin{array}{l} \alpha = (\alpha_1, ..., \alpha_n) \in I\!N^n \\ \beta = (\beta_1, ..., \beta_m) \in I\!N^m \end{array} \right\}$$

forms a k-basis of T. Thus, T forms a quadric solvable polynomial k-algebra with respect to the graded monomial ordering $\succeq_{gr}$ on $\mathcal{B}(T)$, where $\succeq_{gr}$ is defined by extending the ordering $\succeq_{gr}$ on $\mathcal{B}(A)$ and $\mathcal{B}(B)$ to $\mathcal{B}(T)$ in a natural way: If in $\mathcal{B}(A)$, $a_1 \succ_{gr} a_2 \succ_{gr} \cdots \succ_{gr} a_n$, and in $\mathcal{B}(B)$, $b_1 \succ_{gr} b_2 \succ_{gr} \cdots \succ_{gr} b_m$, then set in $\mathcal{B}(T)$

$$X_1 \succ_{gr} X_2 \succ_{gr} \cdots \succ_{gr} X_n \succ_{gr} Y_1 \succ_{gr} Y_2 \succ_{gr} \cdots \succ_{gr} Y_m,$$

and for $\alpha, \gamma \in I\!N^n$, $\beta, \eta \in I\!N^m$, we define $X^\alpha Y^\beta \succ_{gr} X^\gamma Y^\eta$, if

$$\begin{array}{l} \text{either } |\alpha| + |\beta| > |\gamma| + |\eta| \\ \text{or } |\alpha| + |\beta| = |\gamma| + |\eta| \text{ and } \alpha \succ_{gr} \gamma, \\ \text{or } |\alpha| + |\beta| = |\gamma| + |\eta|, \ \alpha = \gamma \text{ and } \beta \succ_{gr} \eta. \end{array}$$

3.1. Lemma With notation as above, if $\mathcal{G}_1 = \{f_1, ..., f_s\}$ is a left Gröbner basis of the left ideal L in A, and $\mathcal{G}_2 = \{g_1, ..., g_h\}$ is a left Gröbner basis of the left ideal K in B, then $\mathcal{G} = \{f_1, ..., f_s, g_1, ..., g_h\}$ is a left Gröbner basis for the left ideal $\mathcal{L} = \langle L, K]$ in T with respect to $\succeq_{gr}$.

Proof Since the $\succeq_{gr}$ on both $\mathcal{B}(A)$ and $\mathcal{B}(B)$ extends to $\mathcal{B}(T)$, one may directly check that every nonzero element in $\mathcal{L}$ has a left Gröbner presentation by $\mathcal{G}$. $\square$

Note that $I\!N^n$ and $I\!N^m$ may be viewed as subsets of $I\!N^{n+m}$ under the natural embeddings

$$\begin{array}{ccc} I\!N^n & \longrightarrow & I\!N^{n+m} \\ (\alpha_1, ..., \alpha_n) & \mapsto & (\alpha_1, ..., \alpha_n, 0, ..., 0) \end{array}$$

and

$$\begin{array}{ccc} I\!N^m & \longrightarrow & I\!N^{n+m} \\ (\beta_1, ..., \beta_m) & \mapsto & (0, ..., 0, \beta_1, ..., \beta_m) \end{array}$$

If we consider the $\succeq_{gr}$-filtrations $\mathcal{F}A$, $\mathcal{F}B$, and $\mathcal{F}T$ respectively (as defined in CH.V §6), it is easy to see that

$$\mathcal{F}_\gamma T = \sum_{\gamma \succeq_{gr} \alpha+\beta} (\mathcal{F}_\alpha A \otimes_k \mathcal{F}_\beta B)\,, \ \gamma \in \mathbb{N}^{n+m}, \ \alpha \in \mathbb{N}^n, \ \beta \in \mathbb{N}^m,$$

$$\frac{\mathcal{F}_\alpha A}{\mathcal{F}_{\alpha^*} A} \bigotimes_k \frac{\mathcal{F}_\beta B}{\mathcal{F}_{\beta^*} B} \cong \frac{\mathcal{F}_\gamma T}{\mathcal{F}_{\gamma^*} T}, \ \gamma \in \mathbb{N}^{n+m}, \ \alpha \in \mathbb{N}^n, \ \beta \in \mathbb{N}^m, \ \alpha+\beta = \gamma.$$

Thus, since $G^\mathcal{F}(A)$ and $G^\mathcal{F}(B)$ are homogeneous solvable polynomial algebras by CH.V Proposition 6.1, we have

$$G^\mathcal{F}(A) \otimes_k G^\mathcal{F}(B) \cong G^\mathcal{F}(T)$$

as *homogeneous* solvable polynomial algebras. Therefore, by the discussion in previous §2 and Lemma 3.1, the following theorem may be obtained by assuming that

(∗) A and B are homogeneous solvable polynomial algebras, and consequently $T = A \otimes_k B$ is a homogeneous solvable polynomial algebra.

3.2. Theorem Let A and B be quadric solvable polynomial algebras and $M = A/L$, $N = B/K$ be as before. We have

$$\begin{aligned} \text{GK.dim}(M \otimes_k N) &= \text{GK.dim}M + \text{GK.dim}N, \\ e(M \otimes_k N) &= e(M)e(N). \end{aligned}$$

Proof We use the above assumption (∗). By Lemma 3.1, we may further assume as in §2 that L and K are *monomial left ideals*, and so $\mathcal{L}$ is a left monomial ideal of T.

Suppose GK.dim$M = p$, $e(M) = \ell$; GK.dim$N = q$, $e(N) = v$. Adopting the notation of §2, by Theorem 2.5 we have

$$\begin{aligned} \mathbf{C}(L) &= C_0 \cup C_1 \cup \cdots \cup C_p \text{ with } C_p \neq \emptyset, \\ h_L(x) &= \frac{\ell}{p!}x^p + \text{lower terms in } x; \\ \mathbf{C}(K) &= C_0 \cup C_1 \cup \cdots \cup C_q \text{ with } C_q \neq \emptyset, \\ h_K(x) &= \frac{v}{q!}x^q + \text{lower terms in } x, \end{aligned}$$

where $h_L(x)$, respectively, $h_K(x)$, is the Hilbert polynomial of M, respectively, of N. We claim that

$$\mathbf{C}(\mathcal{L}) = C_0 \cup C_1 \cup \cdots \cup C_{p+q} \text{ with } C_{p+q} \neq \emptyset.$$

To see this, let us write $e_1, ..., e_n, e_{n+1}, ..., e_{n+m}$ for the unit vectors in $\mathbb{N}^{n+m}$, as defined in §2, and identify $e_1, ..., e_n$ with the unit vectors in $\mathbb{N}^n$, $e_{n+1}, ..., e_{n+m}$

with the unit vectors in $I\!N^m$, where both $I\!N^n$ and $I\!N^m$ are viewed as subsets of $I\!N^{n+m}$. Note that if $\beta + [e_{i_1}, ..., e_{i_r}]$ is a translate of the r-dimensional coordinate subspace $[e_{i_1}, ..., e_{i_r}]$ contained in C_r, then $[e_{i_1}, ..., e_{i_r}]$ is also contained in C_r. Hence, if $[e_{i_1}, ..., e_{i_p}]$ is a p-dimensional coordinate subspace contained in C_p and $[e_{j_1}, ..., e_{j_q}]$ is a q-dimensional coordinate subspace contained in C_q, then by the construction of $\mathcal{L}$ it is easy to see that $[e_{i_1}, ..., e_{i_p}, e_{j_1}, ..., e_{j_q}]$ is a $p+q$-dimensional coordinate subspace contained in C_{p+q}. This shows that $C_{p+q} \neq \emptyset$. Moreover, if $\mathbf{C}(\mathcal{L})$ could contain a $p + q + z$-dimensional coordinate subspace with $z \geq 1$, then again from the construction of $\mathcal{L}$ it is easy to see that $\mathbf{C}(L)$ would contain a $p + z_1$-dimensional coordinate subspace with $z_1 \geq 1$, or $\mathbf{C}(K)$ would contain a $q + z_2$-dimensional coordinate subspace with $z_2 \geq 1$, a contradiction. Therefore, the largest coordinate subspace contained in $\mathbf{C}(\mathcal{L})$ is of dimension $p + q$, and the desired equality GK.dim($M \otimes_k N$) $= p+q =$ GK.dim$M +$ GK.dim$N$ follows from Observation 2.4, the proof of Theorem 2.5(i), and the computation of GK-dimension of modules over homogeneous solvable polynomial algebras in CH.V. In order to prove the equality for the multiplicity, let $\alpha + [e_{i_1}, ..., e_{i_p}]$ be a translate in $C_p \subset \mathbf{C}(L)$, and $\beta + [e_{j_1}, ..., e_{j_q}]$ a translate in $C_q \subset \mathbf{C}(K)$. Then by the construction of $\mathcal{L}$ and the above argumentation we easily see that $(\alpha + \beta) + [e_{i_1}, ..., e_{i_p}, e_{j_1}, ..., e_{j_q}]$ is a translate in $C_{p+q} \subset \mathbf{C}(\mathcal{L})$. Conversely, let $\gamma + [e_{k_1}, ..., e_{k_{p+q}}] \in C_{p+q}$ be a translate of the $p + q$-dimensional coordinate subspace $[e_{k_1}, ..., e_{k_{p+q}}]$, where $\gamma = \sum_{j \notin \{k_1, ..., k_{p+q}\}} s_j e_j$. Again by the above argumentation $[e_{k_1}, ..., e_{k_{p+q}}]$ must be of the form $[e_{i_1}, ..., e_{i_p}, e_{j_1}, ..., e_{j_q}]$ where $[e_{i_1}, ..., e_{i_p}]$ is a p-dimensional coordinate subspace contained in C_p and $[e_{j_1}, ..., e_{j_q}]$ is a q-dimensional coordinate subspace contained in C_q. If we put

$$
\begin{aligned}
J_X &= \left\{ j \notin \{k_1, ..., k_{p+q}\} \;\middle|\; e_j \in \{e_1, ..., e_n\} - \{e_{i_1}, ..., e_{i_p}\} \right\} \\
\alpha &= \sum_{j \in J_X} s_j e_j, \\
J_Y &= \left\{ j \notin \{k_1, ..., k_{p+q}\} \;\middle|\; e_j \in \{e_{n+1}, ..., e_{n+m}\} - \{e_{j_1}, ..., e_{j_q}\} \right\}, \\
\beta &= \sum_{j \in J_Y} h_j e_j,
\end{aligned}
$$

it is easily verified that $\alpha + [e_{i_1}, ..., e_{i_p}]$ is a translate in C_p, $\beta + [e_{j_i}, ..., e_{j_q}]$ is a translate in C_q. Furthermore, it follows from Observation 2.4 and CH.V Theorem 4.2 that $|J_X| = n - p$, $|J_Y| = m - q$. Hence

$$
\gamma + [e_{k_1}, ..., e_{k_{p+q}}] = (\alpha + \beta) + [e_{i_1}, ..., e_{i_p}, e_{j_1}, ..., e_{j_q}].
$$

Thus, we have shown that C_{p+q} contains exactly $\ell v = e(M)e(N)$ distinct translates of the $p + q$-dimensional coordinate subspaces. Therefore, by Theorem 2.5,

$$
e\left(\frac{T}{\mathcal{L}}\right) = ab = e\left(\frac{A}{L}\right) e\left(\frac{B}{K}\right),
$$

as desired. □

Example (i) Let $A = A_1(q) = k[a, b]$ be the additive analogue of the first Weyl algebra subject to the relation $ba = qab + 1$ (CH.I §5 Example (ii)), and let $B = \widetilde{A_1(k)}$ the Rees algebra of the first Weyl algebra $A_1(k)$ with respect to the standard filtration on $A_1(k)$ (CH.I §3). Then by CH.III §3, $B = k[\tilde{a}, \tilde{b}, t]$ is the algebra generated by $\tilde{a}, \tilde{b}, t$ subject to the relations

$$\tilde{a}t = t\tilde{a}, \ \tilde{b}t = t\tilde{b}, \ \tilde{b}\tilde{a} = \tilde{a}\tilde{b} + t^2.$$

Since A is a solvable polynomial algebra with respect to $a >_{grlex} b$, by CH.I §3 and CH.IV §4, B is a quadratic solvable polynomial algebra with respect to $\tilde{a} >_{grlex} \tilde{b} >_{grlex} t$. Let L be the left ideal of A generated by $f = a^2$, $g = ab + (q+1)q^{-2}$. A direct calculation yields $ba^2 = q^2a^2b + (q+1)a$. It turns out that the S-element S(f, g) is zero. Hence $\{f, g\}$ is a left Gröbner basis for L, and GK.dim$(A/L) = 1$, $e(A/L) = 1$. Let K be the left ideal of B generated by $F = \tilde{a}^2$, $G = \tilde{a}\tilde{b} + t^2$. Then since S$(F,G) = H = \tilde{b}F - \tilde{a}G = \tilde{a}t^2$, S$(F, H) = t^2F - \tilde{a}H = 0$, S$(G, H) = t^2G - \tilde{b}H = 0$, it follows that $\{F, G, H\}$ forms a left Gröbner basis for K. By the foregoing results, we read that GK.dim$(B/K) = 2$ and $e(B/K) = 2$. Thus, for the left ideal $\mathcal{L}$ of $T = A \otimes_k B$ generated by L and K, we have GK.dim$(T/\mathcal{L}) = 3$ and $e(T/\mathcal{L}) = 2$.

4. An Application to $A_n(q_1, ..., q_n)$

In view of the first section, the advantage of §3 Theorem 3.2 in the the constructive study of the module representations of a quadric solvable polynomial algebra is obvious, in case the algebra considered is a tensor product of some smaller and nicer quadric solvable polynomial algebras. We illustrate this by an application to the additive analogue $A = A_n(q_1, ..., q_n)$ of the Weyl algebra (CH.I §5 Example (ii)).

Example (i) Recall that $A = k[a_1, ..., a_n, b_1, ..., b_n]$ is defined by the relations

$$\begin{aligned}
a_j a_i &= a_i a_j, \ b_j b_i = b_i b_j, & 1 \le i < j \le n, \\
b_i a_i &= q_i a_i b_i + 1, & i = 1, ..., n, \\
b_j a_i &= a_i b_j, & i \ne j,
\end{aligned}$$

where $q_i \in k - \{0\}$. Using CH.I Theorem 1.1, one may verify that A is the tensor product of its subalgebras $A_1(q_i) = k[a_i, b_i]$, $i = 1, ..., n$, i.e.,

$$A \cong A_1(q_1) \otimes_k A_1(q_2) \otimes_k \cdots \otimes_k A_1(q_n).$$

Put $A(i) = A_1(q_i)$, $i = 1, ..., n$. If $L(i)$ denotes the left ideal of $A(i)$ generated by $\{f_i = a_i^2, \ g_i = a_i b_i + (q_i + 1)q_i^{-2}\}$, and $\mathcal{L}$ denotes the left ideal of A generated by $\{a_i^2, \ a_i b_i + (q_i + 1)q_i^{-2}\}_{i=1}^n$, then by §3 Theorem 3.2 and Example (i), the A-module

$$M = \frac{A}{\mathcal{L}} \cong \frac{A(1)}{L(1)} \otimes_k \frac{A(2)}{L(2)} \otimes_k \cdots \otimes_k \frac{A(n)}{L(n)}$$

has GK.dim$M = n$ and $e(M) = 1$.

Focusing on the (classical) nth Weyl algebra $A_n(k) = k[x_1, ..., x_n, y_1, ..., y_n]$, we characterize and construct the "smallest" modules over $A_n(k)$, algorithmically.

For any nonzero $A_n(k)$-module M, it is well-known that GK.dim$(M) \geq n$ (Bernstein inequality), or in other words, $A_n(k)$ does not have finite dimensional module representation. A nonzero finitely generated $A_n(k)$-module is said to be *holonomic* if GK.dim$(M) = n$ ($= $ gl.dim$A_n(k)$). A holonomic $A_n(k)$-module is *cyclic*, and is n-pure in the sense of §1 (e.g., see [Bor]) and hence is of finite length $\leq e(M)$. If M is a holonomic module with $e(M) = 1$, then M is an irreducible $A_n(k)$-module by Proposition 1.1. So we may say that such modules are the "smallest" modules over $A_n(k)$. Since $A_n(k)$ is a linear solvable polynomial algebra, Theorem 2.11 and CH.V Theorem 7.4 immediately yield the following algorithmic characterization of the smallest $A_n(k)$-modules.

4.1. Theorem Let L be a left ideal of $A_n(k)$ and $\mathcal{G} = \{g_1, ..., g_s\}$ a left Gröbner basis of L with respect to $>_{grlex}$ on $A_n(k)$. Consider the standard filtration on $A_n(k)$ and put

$$
\begin{aligned}
M &= A_n(k)/L, \\
K &= G(L), \\
m_1 &= \mathbf{LM}(\sigma(g_1)), m_2 = \mathbf{LM}(\sigma(g_2)), ..., m_s = \mathbf{LM}(\sigma(g_s)).
\end{aligned}
$$

With notation as in §2, the following statements hold.
(i) GK.dim$(M) = n$ and $e(M) = 1$ if and only if
   (a) $n = \min\{|J| \mid J \in \mathcal{M}\}$; and
   (b) there is only one $J \in \mathcal{M}$ with $J = \{\ell_1, ..., \ell_n\}$ such that $\alpha_{\ell_1} = \alpha_{\ell_2} = \cdots = \alpha_{\ell_n} = 1$, where α_{ℓ_p} is as in §2 (**), $p = 1, ..., n$.
(ii) If GK.dim$(M) = 2n - 1$, then $e(M) = \sum \alpha_i$, where

$$\alpha_i = \min\left\{ \alpha_{ki} \mid \alpha_{ki} \text{ is as in §2 (1)}, \ k = 1, ..., s \right\}, \ i = 1, ..., 2n.$$

$\square$

Here let us point out that in case $n = 1$, the above (i) actually gives more about the generating set of a left ideal in $A_1(k)$.

4.2. Proposition Let L be a left ideal of $A_1(k)$ and $\mathcal{G} = \{g_1, ..., g_s\}$ a Gröbner basis of L with respect to $>_{grlex}$ on $A_1(k)$. Furthermore let α_1, α_2 be as in

Theorem 4.1(ii). If we put $\alpha_1 = \alpha_1$, $\beta_1 = \alpha_2$, and suppose

$$\mathbf{LM}(g_1) = x^{\alpha_1}y^{\beta}, \quad \mathbf{LM}(g_2) = x^{\alpha}y^{\beta_1},$$

then L is generated by $\{g_1, g_2\}$.

Proof Uising the division algorithm in $A_1(k)$, for $f \in L$, if we consider the remainder of f on division by $\{g_1, g_2\}$, it follows from the definition of α_1 and β_1 that L/L' is a finite dimensional k-space, where L' is the left ideal of $A_1(k)$ generated by $\{g_1, g_2\}$. Hence $L = L'$ follows from Bernstein inequality.   □

It is well known (e.g., [Bj]) that every nonzero left ideal of $A_n(k)$ is generated by two elements. Proposition 4.2 may be regarded as an algorithmic realization of this fact in the special $n = 1$ case. We also refer to [Gal] for another algorithmic realization of this special case.

In the foregoing Example (i), if we take all $q_i = 1$, then M becomes a smallest $A_n(k)$-module. Let us finish this chapter by constructing a family of the "smallest" modules over Weyl algebras.

Example (ii) Consider the first Weyl algebra $A_1(k)$ with generators x, y. In [Dix] it is proved that the module $M = A_1(k)/A_1(k)(xy - \beta)$ with $\beta \in k$ is simple if and only if $\beta \notin \mathbb{Z}$. By Theorem 4.1 we see immediately that $e(M) = 2$ because $\{xy - \beta\}$ is a Gröbner basis of the left ideal $A_1(k)(xy - \beta)$ with respect to $x >_{grlex} y$. However, we claim that

• for every integer $n \geq 1$, the module $M = A_1(k)/L$ with L being generated by $\{xy + n, x^n\}$ is a smallest $A_1(k)$-module.

Proof By checking the S-element of $xy + n$ and x^n (CH.II §7) it is easy to see that $\{xy+n, x^n\}$ is a Gröbner basis of L. Hence L is a proper left ideal of $A_1(k)$ and GK.dim$(M) = 1$. From Theorem 4.1 it is also clear that $e(M) = 1$. This shows that M is a smallest $A_1(k)$-module.   □

If we further consider the k-algebra automorphism σ: $A_1(k) \longrightarrow A_1(k)$ with $\sigma(y) = x$, $\sigma(x) = -y$, and the module M in Example (i), then it is easy to see that the *twisted $A_1(k)$-module*, denoted M^{σ}, where $M^{\sigma} \cong M$ as additive groups and the module operation is defined by $fm^{\sigma} = (fm)^{\sigma}$ for $f \in A_1(k)$ and $m^{\sigma} \in M^{\sigma}$, is of the form $A_1(k)/L'$ where L' is the left ideal of $A_1(k)$ generated by $\{xy - (n-1), y^n\}$. It is not hard to verify that M^{σ} is a simple $A_1(k)$-module, and by a similar verification as in Example (i) it is also a simple module of multiplicity 1.

Example (iii) For $n \geq 1$, regarding the subalgebra $A(j)$ of $A_n(k)$ generated by x_j, y_j as the first Weyl algebra, let $L(j)$ be the left ideal of $A(j)$ generated by $\{x_jy_j + j, x_j^j\}$, $j = 1, ..., n$, and K the left ideal of $A_n(k)$ generated by $\{x_jy_j + i,$

$x_j^i\}_{i=j=1}^n$. Then by Example (i)–(ii) and Theorem 3.2,

$$\frac{A_n(k)}{K} \cong \frac{A(1)}{L(1)} \otimes_k \frac{A(2)}{L(2)} \otimes_k \cdots \otimes_k \frac{A(n)}{L(n)}$$

is a simple $A_n(k)$-module with Gelfand-Kirillov dimension n and multiplicity 1.

CHAPTER VII
(∂-)Holonomic Modules and Functions over Quadric Solvable Polynomial Algebras

In this chapter, we show how the noncommutative structure theory is possible to interact with the function identity theory "effectively" via Gröbner bases, in order to support the idea of generalizing Zeilberger's algebraic-algorithmic approach to the automatic proving of holonomic function identities over Weyl algebras [Zei1]. The main point is to clarify how to use the elimination lemma obtained in CH.V §7 in formulating a ∂-holonomicity for both "functions" and modules over certain quadric solvable polynomial algebras (including Weyl algebras), so that automatic proving of multivariate identities over more operator algebras may be carried out by using Gröbner bases. This effectiveness may be illustrated by going back to the work of [Tka3] and [CS] where Gröbner bases were used for eliminating variables and for constructing algorithms such as Creative Telescoping in the automatic proving of multivariate identities. (In the Weyl algebra case the Creative Telescoping is an algorithm written for computing equations satisfied by definite sums or integrals of holonomic functions, e.g., [Zei2]. In [CS] this algorithm has been extended to a relatively wide context.)

To better understand the title of this chapter, we recall in §§2–3 certain relevant notions and background results from the literature. For a remarkable introduction to automatic proving of function identities, we refer to the book: $A = B$, written by M. Petkovšek, H. Wilf and D. Zeilberger, published by A.K. Peters, Ltd. 1996.

Throughout this chapter $k = \mathbb{C}$, the field of complex numbers. Notation for operators adopt those from the literature for the reader's convenience of referring to relevant references.

1. Some Operator Algebras

In this section we list some well-known important operator algebras (e.g., see [CS], [WZ]) that will be used in later sections.

(i) Algebra of linear partial differential operators

As we have pointed out in CH.I §5 and have seen in CH.II §8, the nth Weyl algebra $A_n(k)$ over k coincides with the k-algebra of linear partial differential operators with polynomial coefficients (note that here $k = \mathbb{C}$), i.e., $A_n(k) = \Delta(k[x_1, ..., x_n])$, the subalgebra of the k-algebra $\mathrm{End}_k k[x_1, ..., x_n]$ of all k-linear operators on the commutative polynomial algebra $k[x_1, ..., x_n]$ generated by the operators $\{x_1, ..., x_n, \partial_1, ..., \partial_n\}$, where $\partial_i = \partial/\partial x_i$, $i = 1, ..., n$, and each x_i acts by the left multiplication. The generators of $A_n(k)$ satisfy the following relations.

$$x_i x_j = x_j x_i, \ \partial_i \partial_j = \partial_j \partial_i, \quad 1 \le i < j \le n,$$
$$\partial_j x_i = x_i \partial_j + \delta_{ij}, \qquad\qquad 1 \le i, j \le n.$$

Similarly, let $k(x_1, ..., x_n)$ be the field of rational functions in n variables. Then the k-algebra of linear partial differential operators with rational function coefficients is the algebra $B_n(k) = k(x_1, ..., x_n)[\partial_1, ..., \partial_n] \subset \mathrm{End}_k k(x_1, ..., x_n)$ where the generators satisfy the same relations as that for $A_n(k)$.

(ii) Algebra of linear partial shift operators

The k-algebra of linear partial *shift* (*recurrence*) operators with polynomial coefficients, respectively with rational coefficients, is:

$k[x_1, ..., x_n][E_1, ..., E_m]$, respectively $k(x_1, ..., x_n)[E_1, ..., E_m]$, $n \ge m$, subject to the relations:
$x_j x_i = x_i x_j \ 1 \le i < j \le n,$
$E_i x_i = (x_i + 1)E_i = x_i E_i + E_i, \ 1 \le i \le m,$
$E_j x_i = x_i E_j, \ i \ne j,$
$E_j E_i = E_i E_j, \ 1 \le i < j \le m.$

For any given sequence $u = (u_\alpha)_{\alpha \in \mathbb{N}^n}$ with $\alpha = (\alpha_1, ..., \alpha_i, ..., \alpha_n)$, changing x_i

into integer variable n_i, the actions of n_i and E_i on u are defined as:

$$n_i u = (u_{\alpha'})_{\alpha' \in \mathbb{N}^n} \text{ with } \alpha' = (\alpha_1, ..., \alpha_{i-1}, n_i \alpha_i, \alpha_{i+1}, ..., \alpha_n);$$
$$E_i u = (u_{\alpha''})_{\alpha'' \in \mathbb{N}^n} \text{ with } \alpha'' = (\alpha_1, ..., \alpha_{i-1}, \alpha_i + 1, \alpha_{i+1}, ..., \alpha_n).$$

(iii) Algebra of linear partial difference operators

The k-algebra of linear partial *difference operators* with polynomial coefficients, respectively with rational coefficients, is:

$k[x_1, ..., x_n][\Delta_1, ..., \Delta_m]$, respectively $k(x_1, ..., x_n)[\Delta_1, ..., \Delta_m]$, $n \geq m$, subject to the relations:

$x_j x_i = x_i x_j \ 1 \leq i < j \leq n$,

$\Delta_i x_i = (x_i + 1)\Delta_i + 1 = x_i \Delta_i + \Delta_i + 1, \ 1 \leq i \leq m$,

$\Delta_j x_i = x_i \Delta_j, \ i \neq j$,

$\Delta_j \Delta_i = \Delta_i \Delta_j, \ 1 \leq i < j \leq m$.

For any given function $f = f(x_1, ..., x_n)$, the action of each Δ_i on f is defined as:

$$\Delta_i f = f(x_1, ..., x_{i-1}, x_i + 1, x_{i+1}, ..., x_n) - f.$$

(iv) Algebra of linear partial q-dilation operators

For a fixed $q \in k - \{0\}$, the k-algebra of linear partial q-*dilation operators* with polynomial coefficients, respectively with rational coefficients, is:

$k[x_1, ..., x_n][H_1^{(q)}, ..., H_m^{(q)}]$, respectively $k(x_1, ..., x_n)[H_1^{(q)}, ..., H_m^{(q)}]$, $n \geq m$, subject to the relations:

$x_j x_i = x_i x_j, \ 1 \leq i < j \leq n$,

$H_i^{(q)} x_i = q x_i H_i^{(q)}, \ 1 \leq i \leq m$,

$H_j^{(q)} x_i = x_i H_j^{(q)}, \ i \neq j$,

$H_j^{(q)} H_i^{(q)} = H_i^{(q)} H_j^{(q)}, \ 1 \leq i < j \leq m$.

For any given function $f = f(x_1, ..., x_n)$, the action of each $H_i^{(q)}$ is defined as:

$$H_i^{(q)} f = f(x_1, ..., x_{i-1}, q x_i, x_{i+1}, ..., x_n).$$

(v) Algebra of linear partial q-differential operators

For a fixed $q \in k - \{0\}$, the k-algebra of linear partial q-*differential operators* with polynomial coefficients, respectively with rational coefficients is:

$k[x_1, ..., x_n][D_1^{(q)}, ..., D_m^{(q)}]$, respectively $k(x_1, ..., x_n)[D_1^{(q)}, ..., D_m^{(q)}]$ $n \geq m$, subject to the relations:

$x_j x_i = x_i x_j, \ 1 \leq i < j \leq n$,

$D_i^{(q)} x_i = q x_i D_i^{(q)} + 1, \ 1 \leq i \leq m$,

$D_j^{(q)} x_i = x_i D_j^{(q)}, \ i \neq j$,

$D_j^{(q)} D_i^{(q)} = D_i^{(q)} D_j^{(q)}, \ 1 \leq i < j \leq m$.

For any given function $f = f(x_1, ..., x_n)$, the action of each $D_i^{(q)}$ is defined as:

$$D_i^{(q)} f = \frac{f(x_1, ..., x_{i-1}, qx_i, x_{i+1}, ..., x_n) - f}{(q-1)x_i}.$$

Clearly, if $n = m$, then this operator algebra coincides with the additive analogue of $A_n(k)$ (CH.I §5).

It is not hard to see that all algebras (i) – (v) above are iterated skew polynomial algebras, the algebras of (i), (ii), (iii) and (v) with polynomial coefficients are quadric solvable polynomial algebras, and the algebra of (iv) with polynomial coefficients is a homogeneous solvable polynomial algebra.

2. Holonomic Functions

From both a theoretic and a practical point of view, so far the most successfully machine-proved (found) function identities are those formed by the so called *holonomic functions*. The class of holonomic functions is a subclass of the so called *special functions* which may go back to the earlier book of E.D. Rainville (*Special Functions*, Macmillan, New York, 1960; reprinted: Chelsea, New York, 1971), and may be described as follows. (From the references given below the reader may also trace a short history of automatic proving of function identities.)

- Certain functions appear so often that it is convenient to give them names. These functions are collectively called special functions. There are many examples and no single way of looking at them can illuminate all examples or even all the important properties of a single example of a special function. (— *Special Funtions*, Group Theoretical Aspects and Applications, R.A. Askey, T.H. Koornwinder and W. Schempp, Eds., Reidel, Dordrecht, 1984.)
- Spectial functions have been created and explored to describe scientific and mathematical phenomena. Thrigonometric functions give the relation of angle to length. Riemann's zeta function was invented in order to describe the prime number distribution. Legendre's spherical functions and Bessel's functions were born in connection with the eigenvalue problems for partial differential equations. On the other hand, progress in maths was often made through attempts to understand special functions more deeply. The role of hypergeometric functions in the theory of generalized functions are notable in this respect. Sometimes, developments in maths shed light on previously known special functions from a completely different angle. The rediscover of Painlevé transcendents as correlation functions in statistical mechanical models is one of the best examples of this sort. Several approaches were pursued to   handle various special   functions in a unified manner,   for example, by

means of differential equations, or by group representations. However, the nature of special functions is not yet fully explored. To make further progress, entirely new standpoints are to be sought. The theory of quantum groups applied to functions in q-analogue and the Hodge theory as an abstract theory of algebraic integrals are successful examples of recent innovations. ... (— *Special Functions*, Proc. of the Hayashibara Forum, 1990, Eds. M. Kashiwara and T. Miwa, Springer-Verlag, 1991.)

More about special functions may be found in, e.g., [AW], [Foa], [GR], [GZ].

The study of holonomic function is motivated by a nice fusion of the "continuous" and "discrete" in mathematics, that is the story recalled from (e.g., [CS], [Lip2], [Stan], [Zei1]) as follows.

In view of §1, the nth Weyl algebra $A_n(k)$ is viewed as the differential operator algebra of $k[x_1, ..., x_n]$, i.e, $A_n(k) = k[x_1, ..., x_n][\partial_1, ..., \partial_n] \subset \text{End}_k k[x_1, ..., x_n]$. Since every element of $A_n(k)$ is of the form $D = \sum \lambda_{\alpha\beta} x_1^{\alpha_1} \cdots x_n^{\alpha_n} \partial_1^{\beta_1} \cdots \partial_n^{\beta_n}$, D may be written as a polynomial in $\partial_1, ..., \partial_n$ with coefficients in $k[x_1, ..., x_n]$, i.e., $D = D(\partial_1, ..., \partial_n)$.

Let f be a nonzero member of a family $\mathcal{F}$ on which the Weyl algebra $A_n(k)$ acts naturally. Put

$$\mathcal{I}_f = \left\{ D \subset A_n(k) \mid Df = 0 \right\}$$

Then $\mathcal{I}_f$ is a left ideal of $A_n(k)$ and is called the *annihilator ideal* of f.

(I) C-finite functions
In the case $n = 1$, $A_1(k) = k[x][\partial]$. Recall that the solutions of (homogeneous) linear ordinary differential equations with *constant* coefficients

$$(1) \qquad\qquad D(\partial)f = 0$$

exactly consist of *finite* linear combinations of exponential polynomial solutions, i.e.,

$$f = \sum_{r=1}^{R} P_r(x) e^{\lambda_r x},$$

where $\{\lambda_r\}$ are the roots of the characteristic equation $D(z) = 0$, and the degree of $P_r(x)$, $r = 1, ..., R$, is one less than the multiplicity of the root λ_r. A solution of (1) is called a *C-finite function*.

2.1. Proposition [Zei1] With notation as above, the following are equivalent.
(i) f is C-finite.
(ii) The vector space

$$k[\partial]f = k\text{-Span}\left\{ \partial^i f \mid i \geq 0 \right\}$$

is finite dimensional.

(iii) $k[\partial]/(\mathcal{I}_f \cap k[\partial])$ is a finite dimensional k-space.

(iv) The algebraic set

$$V_f = \left\{ z \in k \ \middle| \ D(z) = 0 \text{ for } D(\partial) \in \mathcal{I}_f \cap k[\partial] \right\}$$

is zero-dimensional (i.e., V_f is finite).

$\square$

In the case where $n > 1$, by the celebrated Ehrenpreis-Palamadov theorem, the solutions of the so called overdetermined systems

(2) $$D_i(\partial_1, ..., \partial_n)f \equiv 0, \quad i = 1, ..., L,$$

with $D_i(\partial_1, ..., \partial_n) \in k[\partial_1, ..., \partial_n]$ exactly consist of (usually infinite) "linear combination" of exponential polynomial functions $P_\lambda(x)e^{\lambda x}$, where $x = (x_1, ..., x_n)$ and $\lambda = (\lambda_1, ..., \lambda_n)$ ranges over the algebraic set

$$V_f = \left\{ \lambda \in k^n \ \middle| \ D_i(\lambda) = 0, \ i = 1, ..., L \right\}.$$

2.2. Proposition [Zei1] With notation as above, the following are equivalent.

(i) f is a solution of the system (2) with a *finite* expression

$$f = \sum_{r=1}^{R} P_r(x)e^{\lambda_r x}.$$

(ii) The vector space

$$k[\partial_1, ..., \partial_n]f = k\text{-Span} \left\{ \partial_1^{\beta_1} \cdots \partial_n^{\beta_n} f \ \middle| \ (\beta_1, ..., \beta_n) \in I\!N^n \right\}$$

is finite dimensional.

(iii) $k[\partial_1, ..., \partial_n]/(\mathcal{I}_f \cap k[\partial_1, ..., \partial_n])$ is a finite dimensional k-space.

(iv) The algebraic set

$$V_f = \left\{ \lambda \in k^n \ \middle| \ D(\lambda) = 0 \text{ for } D(\partial_1, ..., \partial_n) \in \mathcal{I}_f \cap k[\partial_1, ..., \partial_n] \right\}$$

is zero-dimensional (i.e., V_f is finite).

$\square$

A function f satisfying the equivalent statements of Proposition 2.2 is called a *multi-C-finite function*.

Recall from CH.VI §1 that a left $A_n(k)$-module M is called a *holonomic module*, if GK.dim$M = n$ (note that GK.dim$A_n(k) = 2n$). Holonomic module theory over   Weyl algebras was introduced by I.N. Bernstein and M. Kashiwara in the algebraic study of the solutions of linear differential equations (i.e., D-module theory). The earliest grand application of holonomic module theory was given by Bernstein in ([Ber] 1971) for an elementary algebraic proof of a famous conjecture of Gelfand concerning the existence of a meromorphic extension of the distribution valued complex function $\lambda \to P^\lambda$, where P is a polynomial in several variables over $\mathbb{R}^n$.

One of the key properties of modules over $A_n(k)$ is that the *Bernstein inequality* holds:

$$\text{GK.dim}M \geq n \text{ for any nonzero } A_n(k)\text{-module.}$$

Using Bernstein inequality and Proposition 2.2(iii), Zeilberger first observed the following fact.

2.3. Proposition If f is a multi-C-finite function, then the left $A_n(k)$-module $A_n(k)/\mathcal{I}_f$ is holonomic.

□

(II) P-finite functions and sequences

Let $B_n(k) = k(x_1, ..., x_n)[\partial_1, ..., \partial_n]$ be the k-algebra of linear partial differential operators with rational function coefficients (§1 Example (i)). Then $A_n(k)$ is clearly a subalgebra of $B_n(k)$. Let f be a function and

$$\mathcal{J}_f = \left\{ D \in B_n(k) \ \middle| \ Df = 0 \right\},$$

the (left) annihilator ideal of f in $B_n(k)$.

2.4. Proposition ([CS], [Lip1–2], [Stan], [Zei1], a more general case is considered in §4) With notation as above, the following are equivalent.

(i) The family $\{\partial_1^{\alpha_1} \cdots \partial_n^{\alpha_n} f \mid (\alpha_1, ..., \alpha_n) \in \mathbb{N}^n\}$ spans a finite dimensional $k(x_1, ..., x_n)$-space.

(ii) $\mathcal{J}_f \cap k[x_1, ..., x_n][\partial_i] \neq \{0\}$, $i = 1, ..., n$, i.e., f satisfies a system of "ordinary" equations with *polynomial* coefficients:

$$D_i(\partial_i)f = 0, \quad i = 1, ..., n.$$

(iii) The left $B_n(k)$-module $B_n(k)/\mathcal{J}_f$ is a finite dimensional $k(x_1, ..., x_n)$-space.

□

A function f is called *P-finite* if it satisfies the equivalent statements of Proposition 2.4.

Write $\alpha = (\alpha_1, ..., \alpha_n) \in \mathbb{N}^n$, as before. Let $u = (u_\alpha)_{\alpha \in \mathbb{N}^n}$ be a sequence defined on $\mathbb{N}^n$, J a nonempty subset of $\{1, ..., n\}$. For each $i \in J$ and each $a_i \in \mathbb{N}$, define

a. a section of u as any subsequence of u obtained by considering only the terms of u whose indices $\alpha = (\alpha_1, ..., \alpha_n)$ satisfy $\alpha_i = a_i$ for all $i \in J$, i.e., any subsequence obtained by setting at least one index to a given value;

b. an s-section of u as any section of u defined as previously by J and some a_i, with additional constraint that $a_i < s$ for all $i \in J$, where $s \in \mathbb{N}$.

A sequence $u = (u_\alpha)_{\alpha \in \mathbb{N}^n}$ is called *P-recursive* ([Lip2]) if there exists $s \in \mathbb{N}$ such that

(1) for each $i = 1, ..., n$, there exists polynomial $p_\alpha^{(i)}(n_i)$ such that

$$\sum_{\beta \in \{0, ..., s\}^n} p_\beta^{(i)}(\alpha_i) u_{(\alpha - \beta)} = 0$$

when α satisfies $\alpha_i \geq s$ for all $i \in \{1, ..., n\}$;

(2) if $n > 1$ then each s-sections of u satisfies (i) with respect to s.

Let $A = k[x][E]$ be the k-algebra of linear partial shift (recurrence) operators in one variable (see §1). Note that the above definition entails that if $u = (u_\alpha)_{\alpha \in \mathbb{N}}$, then u is P-recursive if and only if there exists $P(E) \in A$ such that

$$P(E)u = 0.$$

In *one* variable case, "P-finite functions" and "P-recursive sequences" were introduced and studied by Stanley in [Stan]. One of the main results of [Stan] is to fuse both notions, as stated below.

2.5. Theorem (P.R. Stanley 1980) A sequence $u = (u_n)_{n \in \mathbb{N}}$ is P-finite if and only if its generating function

$$f(x) = \sum_{n \in \mathbb{N}} u_n x^n$$

is P-finite.

□

In the case of *several* variables a similar result was obtained in [Lip2].

2.6. Theorem (L. Lipshitz 1989) A sequence $u = (u_\alpha)_{\alpha \in \mathbb{N}^n}$ is P-recursive if and only if its generating function

$$f(x_1, ..., x_n) = \sum_{\alpha \in \mathbb{N}^n} u_\alpha x^\alpha, \text{ where } x^\alpha = x_1^{\alpha_1} \cdots x_n^{\alpha_n},$$

is P-finite.

□

Example (i) Legendre polynomials $\{F(n,x)\}_{n\in\mathbb{N}}$ are defined as:

$$F(n,x) = 2^{-n} \sum_{k=0}^{[n/2]} (-1)^k \binom{n}{k} \binom{2(n-k)}{k} x^{n-2k}.$$

Write $\partial F(n,x) = \frac{\partial}{\partial x}F(n,x)$, $EF(n,x) = F(n+1,x)$. Then $F(n,x)$ satisfies a linear differential equation and a linear recurrence equation:

$$\left[(1-x^2)\partial^2 - 2x\partial + n(n+1)\right] F(n,x) = 0$$
$$\left[(n+2)E^2 - (2n+3)xE + (n+1)\right] F(n,x) = 0.$$

Hence $F(n,x)$ is P-finite and $u = \{F(n,x)\}_{n\in\mathbb{N}}$ is P-recursive.

(III) Holonomic functions

The next theorem reveals the relation between holonomic $A_n(k)$-modules and $B_n(k)$-modules which are finite dimensional $k(x_1, ..., x_n)$-spaces. The proof of the "only if" part is given by Bernstein in [Ber] and the proof of the "if" part follows from a result of Kashiwara [Kas] concerning holonomic D-modules (an elementary proof was given by Takayama in [Tak3] but this depends again on the nature of holonomic modules).

2.7. Theorem (Bernstein-Kashiwara) With notation as above, let $\mathcal{J}$ be a left ideal of $B_n(k)$. Then $M = A_n(k)/(\mathcal{J} \cap A_n(k))$ is a holonomic $A_n(k)$-module if and only if $B_n(k)/\mathcal{J}$ is a finite dimensional vector space over $k(x_1, ..., x_n)$.

□

It follows from Proposition 2.4 and Theorem 2.7 that we have the following holonomic module characterization of P-finite functions (sequences).

2.8. Proposition With notation as before, a function f is P-finite if and only if $A_n(k)/(\mathcal{J}_f \cap A_n(k))$ is a holonomic $A_n(k)$-module.

□

Note that $\mathcal{I}_f \subset \mathcal{J}_f$. Bernstein inequality, Proposition 2.8 and Proposition 2.3 immediately yield the following corollary.

2.9. Corollary Any multi-C-finite function f is P-finite.

□

Summing up, the above recalled results lead to the following pure algebraic unification of multi-C-finite function, P-finite function, and P-recursive sequence, as Zeilberger said in [Zei1]:

In the interest of thawing the cold war between the discrete and the
continuous, I have decided to combine these two names into one.

2.10. Definition ([Zei1] 1990) Let f be a nonzero member of a family $\mathcal{F}$ on
which the Weyl algebra $A_n(k)$ acts naturally, and let $\mathcal{I}_f$ be the (left) annihilator
ideal of f in $A_n(k)$. f is said to be a *holonomic function* if the $A_n(k)$-module
$A_n(k)/\mathcal{I}_f$ has Gelfand-Kirillov dimension n, i.e., GK.dim$(A_n(k)/\mathcal{I}_f) = n$, or in
other words, if $A_n(k)/\mathcal{I}_f$ is a holonomic $A_n(k)$-module.

Example (ii) The functions listed below are holonomic (see e.g., [Bj], [CS],
[Lip1–2], [Stan], [WZ], [Zei1]).
(1) Rational functions $1/P$ where P is a nonzero polynomial in $n \geq 1$ variable(s).
(2) All algebraic functions.
(3) All special functions falling in the Askey's scheme (J. Labelle, Tableau
d'Askey, in: *Polynômes Orthogonauxet Applications*, Bar-le-Duc, C. Brezinski
et al., Eds., LNM. 1171, Springer-Verlag, 1985).

3. Automatic Proving of Holonomic Function Identities

Roughly speaking, automatic proving of function identities implies that the following tasks are carried out on computer:

- Find a system of (a finite number of) operator equations satisfied by the given
  function(s) or sequence(s).
- Use the system obtained above and (a finite number of) initial conditions
  to identify the given function(s) or sequence(s) and consequently to prove
  (disprove) the human-posed (human-conjectured) function identities, or to
  find the generating function of a given sequence.
- Find new function identities by performing the operations well-defined on the
  given functions.

Let f be a nonzero member of a family $\mathcal{F}$ on which the Weyl algebra $A_n(k)$
acts naturally. As shown in §2, in the light of Bernstein-Kashiwara theorem for
Weyl algebras, the P-finiteness of f is equivalent to the holonomicity of the left
$A_n(k)$-module $A_n(k)/\mathcal{I}_f$, where $\mathcal{I}_f$ is the (left) annihilator ideal of f in $A_n(k)$.
In this section we describe, briefly, how the holonomicity of modules over Weyl
algebras makes the automatic proving of holonomic function identities feasible,
following [CS], [Tak1–2], [Zei1], and CH.V.

(i) How to verify holonomicity

- For a given function f, find the (left) annihilator ideal $\mathcal{I}_f$ in $A_n(k)$ and produce a Gröbner basis for $\mathcal{I}_f$. This can be directly realized by using one of the Maple packages or one of the computer algebra systems developed by experts in computational algebra (see the websites listed in the introduction).
- Compute the GK-dimension of $A_n(k)/\mathcal{I}_f$ by using the Gröbner basis obtained above. (In the early work of Zeilberger and others there was no systematic algorithmic method for computing GK-dimension. Now we have a quite easy algorithmic method, as shown in CH.V.)

(ii) How to verify holonomic function identities

- The holonomicity is preserved by many algebraic operations, in particular, all holonomic functions form a k-algebra in the usual sense. This big advantage first enables us to recognize many holonomic functions; and moreover, if f and g are holonomic functions, in order to see whether or not f and g are equivalent, it is sufficient to see whether or not $f - g$ is equivalent to 0.
- A holonomic function f can be recovered from a finite amount of information. More precisely, Proposition 2.4(ii), or more generally, the elimination lemma for Weyl algebras (Lemma 3.1 below) guarantees the existence of a "canonical holonomic representation" for f from which it is always possible to know when such a representation is equivalent to 0, and thus it is possible to know when two different canonical representations represent the same function. A canonical holonomic representation can be established by finding n "ordinary" operators

$$(1) \qquad P_i(\partial_i, x_1, ..., x_n), \qquad i = 1, ..., n,$$

in $A_n(k)$ that annihilate f, where each P_i is of degree α_i (in ∂_i), and by fixing the "initial conditions":

$$(2) \qquad \partial_1^{i_1} \cdots \partial_n^{i_n} f(x_0), \quad 0 \le i_1 < \alpha_1, \ ..., \ 0 \le i_n < \alpha_n,$$

where x_0 is any point that is not on the "characteristic set" of the system (1) (the characteristic set of a system (1) is the set of common zeros of the leading coefficients of the operators P_i). This procedure was first recognized by a noncommutative version of Sylvester's dialytic elimination algorithm given by Zeilberger in [Zei1] and later in [WZ] more effectively. Moreover, a good "canonical holonomic representation" may be established by applying left Gröbner bases to the elimination ideals

$$\mathcal{I}_f \cap k[x_1, ..., x_n][\partial_i], \quad i = 1, ..., n,$$

as shown in [CS] and [Tak1–2].

3.1. Lemma (elimination Lemma for Weyl algebras, [Zei1] Lemma 4.1; existence proof is due to I.N. Bernstein) Let L be a left ideal in $A_n(k)$ such that $A_n(k)/L$ is a holonomic $A_n(k)$-module. For every $n+1$ generators selected from the $2n$ generators $\{x_1, ..., x_n, \partial_1, ..., \partial_n\}$ of $A_n(k)$ there is a nonzero member of L that only depends on these $n+1$ generators. In particular, for every $i = 1, ..., n$, L contains a nonzero element of the subalgebra $k[x_1, ..., x_n, \partial_i] \subset A_n(k)$.

□

Elimination lemma for Weyl algebras is nowadays known as the "fundamental lemma" in the automatic proving of holonomic function identities (cf [WZ]). Based on this lemma, effective automatic proving of holonomic function identities has been carried out, and a large class of special function identities including all terminating hypergeometric (alias binomial coefficient) identities has been identified (see "$A = B$"). Using left Gröbner bases, the idea of [Zei1] has been generalized to the automatic proving of multivariate identities in the context of certain iterated Ore extensions (e.g., [CS]), where the holonomicity was replaced by a ∂-finiteness (see §4 for the details).

We finish this section by quoting two simple examples from [Zei1].

Example (i) Let $h = e^{-x} + e^{-x^2}$. A canonical holonomic representation of h is given by

$$D(\partial) = (-2x + 1)\partial^2 + (-4x^2 + 3)\partial + (-4x^2 + 2x + 2),$$
$$h(0) = 2, \ h'(0) = -1.$$

(ii) Let $F(n, x)$ be the Legendre polynomial as in §2 Example (i). A canonical holonomic representation is given by

$$D(\partial) = (1 - x^2)\partial^2 - 2x\partial + n(n - 1),$$
$$F(0, 0) = 1, \ F'(0, 0) = 0, \ F(1, 0) = 0, \ F'(1, 0) = 1.$$

4. Extension/Contraction of the ∂-Finiteness

After seeing how essential the "holonomicity" is in the automatic proving of P-finite function identities, we further explore the Extension/Contraction problem proposed in [CS], in order to have a decent holonomicity over other operator algebras that are different from Weyl algebras. At this stage, the elimination lemma for quadric solvable polynomial algebras (CH.V Lemma 7.6) plays a key role.

To be modest, we restrict our consideration to a class of specific quadric solvable polynomial algebras which we now define, as follows.

Given the sets $\{x_1, ..., x_n\}$, $\{\partial_1, ..., \partial_m\}$ of *symbols* with $n, m \geq 1$, the quadric solvable polynomial algebra $A = k[x_1, ..., x_n, \partial_1, ..., \partial_m]$ we are considering is the one with the associated (left) admissible system $(A, \mathcal{B}, \succeq_{gr})$, where

$$\partial_m \succ_{gr} \partial_{m-1} \succ_{gr} \cdots \succ_{gr} \partial_1 \succ_{gr} x_n \succ_{gr} x_{n-1} \succ_{gr} \cdots \succ_{gr} x_1,$$

$$\mathcal{B} = \left\{ x_1^{\alpha_1} \cdots x_n^{\alpha_n} \partial_1^{\beta_1} \cdots \partial_m^{\beta_m} \;\middle|\; (\alpha_1, ..., \alpha_n, \beta_1, ..., \beta_m) \in \mathbb{N}^{n+m} \right\},$$

and the generators of A satisfy

$$(\diamond) \begin{cases} x_i x_j = x_j x_i, \; 1 \leq i < j \leq n, \\ \partial_\ell x_i = \lambda_{i\ell} x_i \partial_\ell + \sum \lambda_{i\ell}^{p\ell} x_p \partial_\ell + \mathcal{E}_{i\ell} + \lambda, \\ \text{where } \lambda_{i\ell}, \lambda_{i\ell}^{p\ell}, \lambda \in k, \; \lambda_{i\ell} \neq 0, \; 1 \leq i \leq n, \; p < i, \; 1 \leq \ell \leq m, \\ \text{and } \mathcal{E}_{i\ell} \in k\text{-Span}\{x_1, ..., x_n, \partial_\ell\}, \\ \partial_h \partial_\ell = \lambda_{\ell h} \partial_\ell \partial_h + f_{h\ell}, \\ \text{where } 0 \neq \lambda_{\ell h} \in k, \; 1 \leq \ell < h \leq m, \; f_{h\ell} \in A, \; \partial_\ell \partial_h \succ_{gr} \mathbf{LM}(f_{h\ell}). \end{cases}$$

Obviously, A contains the commutative k-algebra $k[x_1, ..., x_n]$ as a subalgebra and every element $D \in A$ may be written as

$$\begin{aligned} D &= \sum c_{(\alpha, \beta)} x_1^{\alpha_1} \cdots x_n^{\alpha_n} \partial_1^{\beta_1} \cdots \partial_m^{\beta_m} \\ &= \sum P_\beta(x_1, ..., x_n) \partial_1^{\beta_1} \cdots \partial_m^{\beta_m}, \end{aligned}$$

where $(\alpha, \beta) = (\alpha_1, ..., \alpha_n, \beta_1, ..., \beta_m) \in \mathbb{N}^{n+m}$ and $P_\beta(x_1, ..., x_n) \in k[x_1, ..., x_n]$. Hence, A can also be written as

$$A = k[x_1, ..., x_n][\partial_1, ..., \partial_m].$$

We further assume that A satisfies the following condition:

(**Q**) $S = k[x_1, ..., x_n] - \{0\}$ forms a left and right Ore set in A, i.e., for any given $s \in S$, $f \in A$, there are $s', s'' \in S$ and $f', f'' \in A$ such that $s'f = f's$ and $fs'' = sf''$.

Except for the Weyl algebra, one easily finds other quadric solvable polynomial algebras satisfying the condition (**Q**), in particular, the algebras listed in §1 with polynomial coefficients are such algebras.

From the condition (**Q**) we know that the localization of A at the Ore set $S = k[x_1, ..., x_n] - \{0\}$, denoted $S^{-1}A$, exists. Thus, we may view A as a subring of $S^{-1}A$ and write $A \subset S^{-1}A$. Bearing this in mind, every element $D \in S^{-1}A$ may be written as

$$D = \sum Q_\beta \partial_1^{\beta_1} \cdots \partial_m^{\beta_m}, \quad Q_\beta \in k(x_1, ..., x_n),$$

where $k(x_1, ..., x_n)$ is the rational function field in variables $x_1, ..., x_n$, i.e., $S^{-1}A$ is indeed a $K(x_1, ..., x_n)$-algebra with generating set $\{\partial_1, ..., \partial_m\}$. So we may write

$$S^{-1}A = k(x_1, ..., x_n)[\partial_1, ..., \partial_m].$$

With the preliminary as above, instead of defining a holonomic-like function, we first introduce a holonomic-like module.

4.1. Definition A left ideal $\mathcal{J}$ of $S^{-1}A$ is said to be ∂-*finite* if $S^{-1}A/\mathcal{J}$ is a finite dimensional vector space over $k(x_1, ..., x_n)$.

4.2. Proposition Let $\mathcal{J}$ be a left ideal of $S^{-1}A$. $\mathcal{J}$ is ∂-finite if and only if for each $i = 1, ..., m$,

$$\mathcal{J} \cap k[x_1, ..., x_n][\partial_i] \neq \{0\}.$$

Proof If $\mathcal{J}$ is ∂-finite, then for each i, $\{1, \partial_i, \partial_i^2, ...\}$ spans a finite dimensional vector space over $k(x_1, ..., x_n)$ in $S^{-1}A/\mathcal{J}$. It follows that there is a nonzero element $P_i(\partial_i) \in k[x_1, ..., x_n][\partial_i] \cap \mathcal{J}$. Conversely, suppose that for each $i = 1, ..., m$ there is a nonzero $P_i(\partial_i) \in k[x_1, ..., x_n][\partial_i] \cap \mathcal{J}$ with degree h_i. Note that since S is an Ore set in A, if $\frac{f}{g} \in k(x_1, ..., x_n)$ and $\alpha_i \geq 1$, then there exist $s \in S$, $a \in A$ such that

$$(1) \qquad s\partial_i^{\alpha_i} = ag \text{ in } A,$$

$$(2) \qquad \frac{\partial_i^{\alpha_i}}{1} \cdot \frac{f}{g} = \frac{af}{s} \text{ in } S^{-1}A.$$

The relations in above ($\diamond$) and (1) above entail that a is a polynomial in ∂_i with coefficients in $k[x_1, ..., x_n]$, and the degree of a with respect to ∂_i is equal to α_i. Thus, the relations in above ($\diamond$) and (2) above entail that $\frac{af}{s}$ is a polynomial in ∂_i with coefficients in $k(x_1, ..., x_n)$ and of degree α_i. Paying attention to the multiplication in $S^{-1}A$, a formal division by $\{P_1(\partial_1), ..., P_m(\partial_m)\}$ in $S^{-1}A$ yields that $S^{-1}A/\mathcal{J}$ is spanned by $\{\partial_1^{\alpha_1} \cdots \partial_m^{\alpha_m}\}_{0 \leq \alpha_i < h_i}$ as a $k(x_1, ..., x_n)$-space. $\square$

Note that the "∂" in the above definition is only a symbol and has no relation with the generators $\partial_1, ..., \partial_m$ of $S^{-1}A$. The name of a ∂-finite ideal comes from [CS] where a class of holonomic-like functions (sequences) was defined by using the ∂-finite ideals (viewed as anihilator ideals of functions) in certain iterated Ore extensions to generalize the work of [Lip1–2], [Tak1–2] and [Zei1–2]. Unfortunately, since a general version of the Bernstein-Kashiwara theorem for general iterated Ore extensions does not exist, the job of finding an "ordinary system", or in other words, the job of eliminating variables in the annihilator ideal, could only be done at the level of manipulating rational function coefficients, i.e., in an algebra of type $S^{-1}A$ instead of in A. This, in turn, led to the extension/contraction problem (see [CS] §4). However, even if one deals only with rational function coefficients, the nice properties of ∂-finite functions (sequences) still yield the relative simplicity of the corresponding algorithm in the automatic proving of (∂-finite) function identities (see [CS] §§1–3).

Let A be an algebra of the type we are considering, and $S^{-1}A$ the corresponding localization of A at $S = k[x_1, ..., x_n] - \{0\}$. If $\mathcal{J}$ is a left ideal of $S^{-1}A$, then the left ideal $\mathcal{J}^c = \mathcal{J} \cap A$ of A is called the *contraction* of $\mathcal{J}$ in A; if L is a left ideal of A, then the left ideal $L^e = S^{-1}AL$ of $S^{-1}A$ is called the *extension* of L in $S^{-1}A$. Note that $\mathcal{J}^{ce} = \mathcal{J}$ and L^{ec} is usually *larger* than L.

Since the ∂-finiteness is defined at the level of $S^{-1}A$, in order to extract the elimination information contained in $\mathcal{J}$ in terms of Gröbner bases, it is natural to pass to the contraction ideal $\mathcal{J}^c$ because A is a quadric solvable polynomial algebra. And it turns out that the next theorem actually provides us a way of dealing with extension/contraction problem with respect to the ∂-finiteness.

For a subset $U = \{x_{i_1}, ..., x_{i_r}, \partial_{j_1}, ..., \partial_{j_t}\} \subset \{x_1, ..., x_n, \partial_1, ..., \partial_m\}$ with $i_1 < i_2 < \cdots < i_r < j_1 < j_2 < \cdots < j_t$, as in CH.V §7 we write $\mathbf{V}(U)$ for the k-vector space spanned by

$$\left\{ x_{i_1}^{\alpha_1} \cdots x_{i_r}^{\alpha_r} \partial_{j_1}^{\beta_1} \cdots \partial_{j_t}^{\beta_t} \;\middle|\; (\alpha_1, ..., \alpha_r, \beta_1, ..., \beta_t) \in I\!N^{r+t} \right\}.$$

4.3. Theorem With notation as before, let L be a left ideal of A, and let $\mathcal{J}$ be a *proper* left ideal of $S^{-1}A$.
(i) Let $U = \{x_{i_1}, ..., x_{i_r}, \partial_{j_1}, ..., \partial_{j_t}\} \subset \{x_1, ..., x_n, \partial_1, ..., \partial_m\}$ be as above. If $L \cap \mathbf{V}(U) = \{0\}$, then $GK.\dim(A/L) \geq r + t$.
(ii) If $GK.\dim(A/L) = n$, then for each $i = 1, ..., m$, $U = \{x_1, ..., x_n, \partial_i\}$,

$$L \cap \mathbf{V}(U) \neq \{0\}.$$

Hence, the extension left ideal L^e of L in $S^{-1}A$ is ∂-finite.
(iii) $GK.\dim(A/\mathcal{J}^c) \geq n$.
(iv) If $GK.\dim(A/L) = n$ and $L \cap k[x_1, ..., x_n] = \{0\}$, then

$$GK.\dim(A/L^{ec}) = n.$$

Consequently, if $GK.\dim(A/\mathcal{J}^c) = n$ then $\mathcal{J}$ is ∂-finite in $S^{-1}A$.

Proof (i) and (ii) follow from CH.V Theorem 5.7 (or its version for general quadric solvable polynomial algebras) and Lemma 7.5 (elimination lemma). Since $\mathcal{J}$ is a proper left ideal of $S^{-1}A$ and the localization (or the extension) of the contraction left ideal $\mathcal{J}^c$ is equal to $\mathcal{J}$, it is clear that $\mathcal{J}^c \cap k[x_1, ..., x_n] = \{0\}$. Hence (iii) follows from (i). Finally, (iv) follows from (ii), (iii) and the natural A-module epimorphism $A/L \rightarrow A/L^{ec}$ because L^e is now a proper left ideal of $S^{-1}A$ by the assumption on L.   □

5. The ∂-Holonomicity

Let A be a quadric solvable polynomial algebra as in last section. By the discussion of previous sections, we are now ready to introduce the ∂-holonomicity for both "functions" and modules over A. Notation are maintained as before.

Let $S^{-1}A$ be the localization of A at $S = k[x_1, ..., x_n] - \{0\}$, and let $\mathcal{F}$ be a family of "functions" including $k[x_1, ..., x_n]$. Suppose that $S^{-1}A$ (and hence A) acts on $\mathcal{F}$ naturally, or that $\mathcal{F}$ forms a left $S^{-1}A$-module (hence an A-module). For a nonzero member $f \in \mathcal{F}$, we consider the (left) annihilator ideal of f in $S^{-1}A$, respectively in A:

$$\mathcal{J}_f = \left\{ D \in S^{-1}A \mid Df = 0 \right\},$$

$$\mathcal{I}_f = \left\{ D \in A \mid Df = 0 \right\}.$$

It is clear that $\mathcal{J}_f$ is a *proper* left ideal of $S^{-1}A$, $\mathcal{J}_f \cap k[x_1, ..., x_n] = \{0\}$, and $\mathcal{I}_f \cap k[x_1, ..., x_n] = \{0\}$. In particular, since $\mathcal{J}_f^c = \mathcal{J}_f \cap A = \mathcal{I}_f$, it follows from Theorem 4.3 that

$$\text{GK.dim}(A/\mathcal{J}_f^c) = \text{GK.dim}(A/\mathcal{I}_f) \geq n.$$

In view of Theorem 4.3 and the above remark, it is natural to define ∂-finite functions, ∂-holonomic functions, ∂-finite modules and ∂-holonomic modules as follows.

5.1. Definition With notation as above, let f be a nonzero member of $\mathcal{F}$, and let M be a finitely generated A-module.
(i) f is said to be a *∂-finite function* if $\mathcal{J}_f$ is a ∂-finite left ideal of $S^{-1}A$ in the sense of Definition 4.1, i.e., $S^{-1}A/\mathcal{J}_f$ is a finite dimensional $k(x_1, ..., x_n)$-space.
(ii) f is said to be a *∂-holonomic function* if $\text{GK.dim}(A/\mathcal{I}_f) = n$.
(iii) M is called a *∂-finite A-module* if the localization of M at S, denoted $S^{-1}M$, is finite dimensional over $k(x_1, ..., x_n)$.
(iv) M is called *∂-holonomic* if $\text{GK.dim}M = n$.

Remark (i) (compare with Definition 2.10.) From Theorem 4.3 and Definition 5.1 it is clear that the ∂-holonomicity of f implies the ∂-finiteness of f; but the converse is not necessarily true for algebras of type A (the author believes this but failed to have a counterexample). Some discussion on the equivalence of ∂-holonomicity and ∂-finiteness is given in connection with the generalization of Bernstein-kashiwara theorem (Theorem 2.7) after Proposition 5.2 below.
(ii) Since for a left ideal L of a quadric solvable polynomial algebra A the Gelfand-Kirillov dimension of the A-module A/L is always computable in terms of Gröbner basis (CH.V), we see that the ∂-holonomicity defined above is algorithmically recognizable.

Next we show that the automatic proving of ∂-finite function identities, respectively of ∂-holonomic function identities, is feasible in the following sense: if f and g are ∂-finite functions, respectively ∂-holonomic functions such that $f - g$ is the same type of function, then, to prove f and g are the same function, it is sufficient to show that $f - g$ is equivalent to the zero function by using the ∂-finiteness, respectively the ∂-holonomicity, as one did in the context of Weyl algebras. In particular, the automatic proving of ∂-holonomic function identities can be done at the level of manipulating polynomial function coefficients.

5.2. Proposition With notation as above, let $f,\ g \in \mathcal{F}$ be ∂-finite, respectively ∂-holonomic functions in the sense of definition 5.1. Then $f + g$ is a ∂-finite, respectively a ∂-holonomic function.

Proof If $f, g \in \mathcal{F}$ are ∂-finite, respectively ∂-holonomic functions, then we have the $S^{-1}A$-module $S^{-1}Af \oplus S^{-1}Ag$, respectively the A-module $Af \oplus Ag$. Considering the element (f, g) in both modules and the $S^{-1}A$-submodule $S^{-1}A(f, g)$, respectively the A-submodule $A(f, g)$, we obtain the following exact sequences of modules:

$$0 \longrightarrow \frac{S^{-1}A}{\mathcal{J}_{(f,g)}} \longrightarrow \frac{S^{-1}A}{\mathcal{J}_f} \oplus \frac{S^{-1}A}{\mathcal{J}_g}$$

$$\frac{S^{-1}A}{\mathcal{J}_{(f,g)}} \longrightarrow \frac{S^{-1}A}{\mathcal{J}_{f+g}} \longrightarrow 0$$

$$0 \longrightarrow \frac{A}{\mathcal{I}_{(f,g)}} \longrightarrow \frac{A}{\mathcal{I}_f} \oplus \frac{A}{\mathcal{I}_g}$$

$$\frac{A}{\mathcal{I}_{(f,g)}} \longrightarrow \frac{A}{\mathcal{I}_{f+g}} \longrightarrow 0$$

where $\mathcal{J}_{(f,g)}$, respectively $\mathcal{I}_{(f,g)}$ is the (left) annihilator ideal of (f, g) in $S^{-1}A$, respectively in A. Since

$$\dim_{k(x_1,\dots,x_n)} \left(\frac{S^{-1}A}{\mathcal{J}_f} \oplus \frac{S^{-1}A}{\mathcal{J}_g} \right) < \infty, \quad \mathrm{GK.dim} \left(\frac{A}{\mathcal{I}_f} \oplus \frac{A}{\mathcal{I}_g} \right) = n,$$

it follows from Theorem 4.3(iii) that:

$$\dim_{k(x_1,\dots,x_n)} \frac{S^{-1}A}{\mathcal{I}_{f+g}} < \infty, \quad \mathrm{GK.dim} \frac{A}{\mathcal{I}_{f+g}} = n,$$

as desired. $\qquad\qquad\square$

Finally, we discuss the possibility of extending the Bernstein-Kashiwara theorem to algebras of type A.

It follows from the foregoing remark that if the ∂-finiteness always implied ∂-holonomicity, then we might replace ∂-finiteness by ∂-holonomicity, and consequently the automatic proving of ∂-finite function identities could be reduced completely from manipulating rational function coefficients to manipulating polynomial function coefficients. More clearly, we might expect that the equality in Theorem 4.3(iii) holds, and hence we would have an analogue of the Bernstein-Kashiwara theorem for algebras of type A. By CH.V Theorem 7.4 the most obvious case where the equality $\mathrm{GK.dim}(A/\mathcal{J}^c) = n$ holds is when $\mathcal{J}^c$ contains elements of the form $\partial_j^{\beta_j} + f_j$ of A, where the f_j are elements of A with $\partial_j^{\beta_j} = \mathbf{LM}(\partial_j^{\beta_j}) \succ_{gr} \mathbf{LM}(f_j), j = 1, ..., m$. However, it seems to the author that the equality of Theorem 4.3(iii) does not hold for an arbitrary ∂-finite left ideal in $S^{-1}A$. Nevertheless, we prove that for certain algebras of type A (different from the Weyl algebras) the equality in Proposition 4.3(iii) may hold for arbitrary ∂-finite left ideals of $S^{-1}A$.

5.3. Proposition If A is generated by $n + 1$ elements $x_1, ..., x_n, \partial$, i.e., $A = k[x_1, ..., x_n, \partial]$, and $\mathcal{J}$ is any proper left ideal of $S^{-1}A$, then $\mathrm{GK.dim}(A/\mathcal{J}^c) = n$. Moreover, every proper left ideal in $S^{-1}A$ if ∂-finite.

Proof That $\mathrm{GK.dim}(A/\mathcal{J}^c) = n$ follows from Theorem 4.3 and CH.V Theorem 5.7. The ∂-finiteness of $\mathcal{J}$ follows from a formal division as in the proof of Proposition 4.2 by considering the polynomial $P(\partial)$ in ∂ with coefficients in $k[x_1, ..., x_n]$ which is contained in $\mathcal{J}$ and has the smallest degree.   □

Another example is obtained by considering the algebra $A = k[x_1, ..., x_n, \partial_1, ..., \partial_m]$ with $n, m > 1$ but we assume that in $S^{-1}A$:

$$
(*) \qquad \begin{cases} \partial_j \cdot \dfrac{1}{f^\ell} = \dfrac{f^\ell \partial_j - \delta(f^\ell)}{f^{2\ell}}, \quad j = 1, ..., m, \ \ell > 0, \\[3mm] \text{where } f, \delta(f) \in k[x_1, ..., x_n] \text{ with } \deg\delta(f) \le \deg f. \end{cases}
$$

It is not difficult to find such algebras (including Weyl algebras as a special case).

5.4. Proposition Let A be a k-algebra as above and $S^{-1}A$ the localization of A at $S = k[x_1, ..., x_n] - \{0\}$. If $\mathcal{J}$ is a ∂-finite proper left ideal of $S^{-1}A$, then $\mathrm{GK.dim}(A/\mathcal{J}^c) = n$.

Proof By Theorem 4.3(iii), we only have to show that $\mathrm{GK.dim}(A/\mathcal{J}^c) \le n$. Let $\overline{s_1}, ..., \overline{s_t}$ be a basis of the $k(x_1, ..., x_n)$-space $S^{-1}A/\mathcal{J}$. We may assume that the $\overline{s_j}$ are classes of the monomials $s_j = \partial_1^{\beta_{j1}} \cdots \partial_m^{\beta_{jm}}$ in $S^{-1}A/\mathcal{J}$ and that $s_1 = 1$. Since $\{\overline{s_1} = 1, \overline{s_2}, ..., \overline{s_t}\}$ is a basis, there exists a $p \in k[x_1, ..., x_n]$ and

$q_{vj}^u \in k[x_1, ..., x_n]$, $1 \leq u \leq m$, $1 \leq v, j \leq t$, such that

$$(**) \qquad\qquad p\overline{\partial_u s_v} = \sum_{j=1}^{t} q_{vj}^u \overline{s_j}.$$

Let p be the polynomial we fixed in $(**)$, and consider the k-subspace M of $S^{-1}A/\mathcal{J}$ which is defined as follows:

$$M = \sum_{j=1}^{t} k[x_1, ..., x_n, p^{-1}]\overline{s_j}.$$

Putting

$$E = \max_{1 \leq u \leq m,\; 1 \leq v, j \leq t} \left\{ \deg p + 1, \; \deg q_{vj}^u \right\},$$

$$T = 2E,$$

M has a filtration consisting of k-subspaces:

$$F_w M = \left\{ p^{-w} \sum_{j=1}^{t} g_j s_j \;\middle|\; g_j \in k[x_1, ..., x_n], \; \deg g_j \leq wT \right\}, \quad w \geq 0,$$

and moreover $\dim_k F_w M = t \cdot \begin{pmatrix} wT + n \\ n \end{pmatrix}$ which is a polynomial in w of degree n. If we consider the filtration on $(A + \mathcal{J})/\mathcal{J} \cong A/\mathcal{J}^c$ induced by the standard filtration FA on A, it follows from CH.V Theorem 5.2 and the formula (∇) before CH.V Corollary 5.4 (or their version for general quadric solvable polynomial algebras) that we can finish the proof by showing that

$$F_w(A/\mathcal{J}^c) \subset F_{2^w} M, \quad w \geq 0.$$

Indeed, let $D = x_1^{\alpha_1} \cdots x_n^{\alpha_n} \overline{\partial_1^{\beta_1} \cdots \partial_m^{\beta_m}}$ be a monomial in $A/\mathcal{J}^c$ such that $|\alpha| + |\beta| \leq w$, where $|\alpha| = \alpha_1 + \cdots + \alpha_n$, $|\beta| = \beta_1 + \cdots + \beta_m$. If we start with $\overline{\partial_m}$ in the foregoing $(**)$:

$$p\overline{\partial_m} = p\overline{\partial_m s_1} = \sum_{j=1}^{t} q_{1j}^m \overline{s_j} \text{ implying } \overline{\partial_m} = \frac{1}{p}\sum_{j=1}^{t} q_{1j}^m \overline{s_j},$$

then by the assumption $(*)$ we obtain

$$
\begin{aligned}
\overline{\partial_m^2} = \partial_m \overline{\partial_m} \;&=\; \partial_m \cdot \frac{1}{p}\sum_{j=1}^{t} q_{1j}^m \overline{s_j} \\[2ex]
&=\; \frac{1}{p^2}\sum_{j=1}^{t} \left(p\partial_m - \delta(p)\right) q_{1j}^m \overline{s_j} \\[2ex]
&=\; \frac{1}{p^2}\sum_{j=1}^{t} \left(p\left(q_{1j}^m \partial_m + \delta(q_{1j}^m)\right)\overline{s_j} + \delta(p)\overline{s_j}\right) \\[2ex]
&=\; \frac{1}{p^2}\sum_{j=1}^{t} \left(q_{1j}^m p\overline{\partial_m s_j} + p\delta(q_{1j}^m)\overline{s_j} + \delta(p)\overline{s_j}\right) \\[2ex]
&=\; \frac{1}{p^2}\sum_{j=1}^{t} \left(q_{1j}^m \left(\sum_{h=1}^{t} q_{jh}^m \overline{s_h}\right) + p\delta(q_{1j}^m)\overline{s_j} + \delta(p)\overline{s_j}\right) \\[2ex]
&=\; \frac{1}{p^2}\sum_{j=1}^{t} q_j \overline{s_j}
\end{aligned}
$$

with $q_j \in k[x_1,...,x_n]$ and $\deg q_j \le 2E$. A repetition of this procedure yields:

$$
\overline{\partial_m^{\beta_m}} = \partial_m \overline{\partial_m^{\beta_m-1}} = \frac{1}{p^{2^{\beta_m}-1}}\sum_{j=1}^{t} q_j \overline{s_j} \text{ with } q_j \in k[x_1,...,x_n], \; \deg q_j \le 2^{\beta_m-1}\cdot E,
$$

$$
\vdots
$$

$$
\overline{\partial_1^{\beta_1}\cdots\partial_m^{\beta_m}} = \frac{1}{p^{2^{|\beta|-1}}}\sum_{j=1}^{t} q_j' \overline{s_j} \text{ with } q_j' \in k[x_1,...,x_n], \; \deg q_j' \le 2^{|\beta|-1}\cdot E,
$$

and hence

$$
\begin{aligned}
D = x_1^{\alpha_1}\cdots x_n^{\alpha_n}\overline{\partial_1^{\beta_1}\cdots\partial_m^{\beta_m}} \;&=\; \frac{1}{p^{2^{|\beta|-1}}}\sum_{j=1}^{t} x_1^{\alpha_1}\cdots x_n^{\alpha_n} q_j' \overline{s_j} \\[2ex]
&=\; \frac{1}{p^{2^{|\alpha|+|\beta|}}}\sum_{j=1}^{t} p^{2^{|\alpha|+|\beta|}-2^{|\beta|-1}} x_1^{\alpha_1}\cdots x_n^{\alpha_n} q_j' \overline{s_j}.
\end{aligned}
$$

Here

$$\deg\left(p^{2^{|\alpha|+|\beta|}-2^{|\beta|-1}} x_1^{\alpha_1}\cdots x_n^{\alpha_n} q_j'\right) \begin{array}{l} \leq \ (2^{|\alpha|+|\beta|}-2^{|\beta|-1})\deg p + |\alpha| + 2^{|\beta|-1}\cdot E \\ \leq \ (2^{|\alpha|+|\beta|}-2^{|\beta|-1})\deg p + 2^{|\alpha|} + 2^{|\beta|-1}\cdot E \\ \leq \ 2^{|\alpha|+|\beta|}\cdot E + \left(2^{|\alpha|}+2^{|\beta|}\right)\cdot E \\ \leq \ 2\cdot 2^{|\alpha|+|\beta|}\cdot E \\ = \ 2^{|\alpha|+|\beta|}\cdot T \\ \leq \ 2^w \cdot T. \end{array}$$

This shows that $D \in F_{|\alpha|+|\beta|}M \subset F_{2^w}M$, as desired. $\square$.

CHAPTER VIII
Regularity and K_0-group of Quadric Solvable Polynomial Algebras

From CH.III §2 we have seen that the class of quadric solvable polynomial algebras includes many popular algebras that are Auslander regular (see CH.VI §1 for the definition) and have K_0-group $\mathbb{Z}$ (hence every finitely generated module has a finite free resolution, or in other words, a noncommutative version of Hilbert's syzygy theorem holds), e.g., the Weyl algebra and its additive and multiplicative analogues, enveloping algebras of finite dimensional Lie algebras, q-Heisenberg algebras, ... To qualify the name of a quadric solvable polynomial algebra A (comparing with a commutative polynomial algebra), the results obtained in previous chapters naturally motivate us further to explore the regularity of A (at least at the level of having finite global dimension and K_0-group $\mathbb{Z}$). In the case where a given quadric solvable polynomial algebra A is an iterated skew polynomial algebra starting with the ground field k, the regularity and K_0-group of A are well-known. However, enveloping algebras of Lie algebras and examples constructed in CH.III §2 show that quadric solvable polynomial algebras are not always iterated skew polynomial algebras starting with the ground field k. In this final chapter, we first derive that every *tame* quadric solvable polynomial algebra A is Auslander regular with $K_0(A) \cong \mathbb{Z}$. This is achieved by a closer look at the associated graded algebra $G(A)$ of A with respect to the standard filtration FA. After introducing the $\succeq_{gr}$-filtration on modules, we prove, by passing to the associated $I\!N^n$-graded algebra $G^{\mathcal{F}}(A)$ of A with respect to its $\succeq_{gr}$-filtration $\mathcal{F}A$, that every quadric solvable polynomial algebra A is of finite global dimension. Backing to the standard filtration again in the final section, it is proved that $K_0(A) \cong \mathbb{Z}$ holds for every quadric solvable polynomial algebra A. At this stage, we may say that every quadric solvable polynomial algebra is regular in the classical sense. Yet, the author strongly believes that the following proposition is true, though he himself failed to prove it in general.

Proposition Every quadric solvable polynomial algebra A is Auslander regular.

Note that since every quadric solvable polynomial algebra is a left and right Noetherian domain (with 1) over a field, the invariant basis property holds for such algebras, and consequently, there is no problem to talk about global dimension and K_0-group of such algebras. We refer to [Rot] for general homological algebra, and refer to [Bas] for general algebraic K-theory.

1. Tame Case: A is Auslander Regular with $K_0(A) \cong \mathbb{Z}$

Let $A = k[a_1, ..., a_n]$ be a *tame* quadric solvable polynomial algebra in the sense of CH.III Definition 2.1, and $(A, \mathcal{B}, \succeq_{gr})$ the associated (left) admissible system, where

$$\mathcal{B} = \{a_1^{\alpha_1} a_2^{\alpha_2} \cdots a_n^{\alpha_n} \mid (\alpha_1, ..., \alpha_n) \in I\!N^n\},$$

and $\succeq_{gr}$ is a graded monomial ordering on $\mathcal{B}$. Then it follows from CH.III Proposition 2.3 that A is completely constructable. In this section, we prove that A and its associated graded structures with respect to the standard filtration FA on A are Auslander regular with K_0-group $\mathbb{Z}$. Notation is retained as in previous chapters.

1.1. Theorem Let A be as above, and let $G(A)$ and $\widetilde{A}$ be the associated graded algebra and Rees algebra of A with respect to the standard filtration FA on A, respectively. Then A, $G(A)$, and $\widetilde{A}$ are Auslander regular domains with K_0-group $\mathbb{Z}$.

Proof By the definition of a tame quadric solvable polynomial algebra and CH.IV Theorem 4.1, the generators of A satisfy

$$a_j a_i = \lambda_{ji} a_i a_j + \sum_{k \leq \ell < j} \lambda_{ji}^{k\ell} a_k a_\ell + \sum \lambda_h a_h + c_{ji}, \ 1 \leq i < j \leq n,$$

where $\lambda_{ji} \neq 0$, $\lambda_{ji}^{k\ell}$, λ_h, $c_{ji} \in k$, the generators of $G(A) = k[\sigma(a_1), ..., \sigma(a_n)]$ satisfy the quadratic relations

$$\sigma(a_j)\sigma(a_i) = \lambda_{ji}\sigma(a_i)\sigma(a_j) + \sum_{k,\ell<j} \lambda_{ji}^{k\ell}\sigma(a_k)\sigma(a_\ell), \ 1 \leq i < j \leq n,$$

and $G(A)$ has the k-basis

$$\sigma(\mathcal{B}) = \left\{ \sigma(a_1)^{\alpha_1}\sigma(a_2)^{\alpha_2}\cdots\sigma(a_n)^{\alpha_n} \ \middle| \ (\alpha_1, ..., \alpha_n) \in I\!N^n \right\}.$$

Consequently, the above defining relations determine an iterated skew polyno-mial algebra structure starting with the polynomial algebra $k[\sigma(x_1)]$. Therefore, $G(A)$ is an Auslander regular domain. It follows from [LVO1] and [LVO2] that A and $\widetilde{A}$ are Auslander regular domains, and it follows from the K_0-part of Quillen's theorem ([Qui] Theorem 7) that

$$
\begin{array}{ccc}
 & K_0(F_0A) & \cong & K_0(A) \\
 & & =\nearrow & \\
\mathbb{Z} \xrightarrow{\cong} K_0(k) & \xrightarrow{=} K_0(G(A)_0) & \cong & K_0(G(A)) \\
 & & =\searrow & \\
 & K_0(\widetilde{A}_0) & \cong & K_0(\widetilde{A})
\end{array}
$$

as desired.   □

2. The $\succeq_{gr}$-filtration on Modules

From CH.III Definition 2.1 and the examples constructed in CH.III §2 we know that not every quadric solvable polynommial algebra is a tame quadric solv-able polynomial algebra. To study the regularity and K_0-group of an arbitrary quadric solvable polynomial algebra $A = k[a_1, ..., a_n]$, in this section we introduce the $\succeq_{gr}$-filtration on A-modules and discuss the $\succeq_{gr}$-filtered homomorphisms and the associated $I\!N^n$-graded homomorphisms.

Let $(A, \mathcal{B}, \succeq_{gr})$ be the (left) admissible system associated with A, where

$$\mathcal{B} = \left\{ a^\alpha = a_1^{\alpha_1} \ldots a_n^{\alpha_n} \ \middle| \ \alpha = (\alpha_1, ..., \alpha_n) \in I\!N^n \right\},$$

and $\succeq_{gr}$ is a graded monomial ordering on $\mathcal{B}$. Let $\mathcal{F}A$ be the $\succeq_{gr}$-filtration $\mathcal{F}A$ on A and $G^{\mathcal{F}}(A)$ the associated $I\!N^n$-graded algebra of A with respect to $\mathcal{F}A$, as defined in CH.V §6.

2.1. Definition Let M be a (left) A-module. M is said to be a $\succeq_{gr}$-*filtered A-module* if there is a family $\mathcal{F}M = \{\mathcal{F}_\alpha M\}_{\alpha \in I\!N^n}$ consisting of k-subspaces $\mathcal{F}_\alpha M$ of M such that
   $(\mathcal{F}M1)$ $\cup_{\alpha \in I\!N^n} \mathcal{F}_\alpha M = M$,
   $(\mathcal{F}M2)$ $\mathcal{F}_\beta M \subset \mathcal{F}_\alpha M$ if $\alpha \succeq_{gr} \beta$, and
   $(\mathcal{F}M3)$ $(\mathcal{F}_\alpha A)(\mathcal{F}_\beta M) \subset \mathcal{F}_{\alpha+\beta} M$ for all $\alpha, \beta \in I\!N^n$.
$\mathcal{F}M$ is called a $\succeq_{gr}$-*filtration* on M.
Since for each $\alpha \in I\!N^n$ there is $\alpha^* = \max\{\gamma \in I\!N^n \mid \alpha \succ_{gr} \gamma\}$, to be convenient, for the least element $0 = (0, ..., 0) \in I\!N^n$, we set $\mathcal{F}_{0^*} M = \{0\}$ in every $\mathcal{F}M$.

If M is a $\succeq_{gr}$-filtered A-module with $\succeq_{gr}$-filtration $\mathcal{F}M$, then the *associated graded $G^{\mathcal{F}}(A)$-modules* of M is defined as the $I\!N^n$-graded additive group

$$G^{\mathcal{F}}(M) = \bigoplus_{\alpha \in I\!N^n} G^{\mathcal{F}}(M)_\alpha \text{ with } G^{\mathcal{F}}(M)_\alpha = \mathcal{F}_\alpha M / \mathcal{F}_{\alpha^*} M$$

on which the module action of $G^{\mathcal{F}}(A)$ is given by

$$G^{\mathcal{F}}(A)_\alpha \times G^{\mathcal{F}}(M)_\beta \longrightarrow G^{\mathcal{F}}(M)_{\alpha+\beta}$$

$$(\overline{f}, \overline{m}) \longmapsto \overline{fm}$$

where, if $f \in \mathcal{F}_\alpha A$, respectively if $m \in \mathcal{F}_\beta M$, then $\overline{f}$ stands for the image of $f \in G^{\mathcal{F}}(A)_\alpha = \mathcal{F}_\alpha A / \mathcal{F}_{\alpha^*} A$, respectively $\overline{m}$ stands for the image of m in $G^{\mathcal{F}}(M)_\beta = \mathcal{F}_\beta M / \mathcal{F}_{\beta^*} M$.
A $\succeq_{gr}$-filtration $\mathcal{F}M$ has the property that if $0 \neq m \in M$, then there is $\alpha \in I\!N^n$ such that $m \in \mathcal{F}_\alpha M - \mathcal{F}_{\alpha^*} M$. In this case we call α the *degree* of m and write $\sigma(m)$ for its corresponding homogeneous element in $G^{\mathcal{F}}(M)_\alpha$.

Before dealing with the associated $I\!N^n$-graded $G^{\mathcal{F}}(A)$-module $G^{\mathcal{F}}(M)$ of a $\succeq_{gr}$-filtered A-module M with $\succeq_{gr}$-filtration $\mathcal{F}M$, we first note that, for $\alpha, \beta \in I\!N^n$ with $\alpha \succ_{gr} \beta$, the equation $\alpha = \beta + x$ does not necessarily have a solution in $I\!N^n$. In particular, even if for $\alpha \succ_{gr} \beta$ and $\alpha^* \succ_{gr} \beta$, by the definition of α^*, the equations

$$\alpha = \beta + x \text{ and } \alpha^* = \beta + y$$

may not have solutions in $I\!N^n$ simultaneously. This makes the $\succeq_{gr}$-filtrations behave quite different from $\mathbb{Z}$-filtrations. To remedy this defect, let us put

$$[0, \alpha] = \Big\{ \gamma \in I\!N^n \mid \alpha \succeq_{gr} \gamma \Big\}.$$

Then clearly, $\alpha^* = \max\{[0, \alpha] - \{\alpha\}\}$.

2.2. Lemma Let $\alpha, \eta \in I\!N^n$ be such that $\alpha = \eta + \gamma$ for some $\gamma \in I\!N^n$. For any $\beta \in [0, \alpha^*]$, if $\beta = \eta + \delta$ for some $\delta \in I\!N^n$, then $\gamma^* \succeq_{gr} \delta$; and if $\beta = \alpha^*$, then $\delta = \gamma^*$.

Proof Note that $\succeq_{gr}$ is a monomial ordering. The first conclusion is then clear by the definition of a $*$-element. Suppose $\alpha^* = \eta + \delta$. Then, $\gamma \succ_{gr} \gamma^*$ implies $\alpha = \eta + \gamma \succ_{gr} \eta + \gamma^*$. This, in turn, implies $\eta + \delta = \alpha^* \succeq_{gr} \eta + \gamma^*$, and hence $\delta \succeq_{gr} \gamma^*$. Combining the first conclusion, we conclude that $\delta = \gamma^*$.   □

2.3. Proposition Let M be an A-module.
(i) If M has a $\succeq_{gr}$-filtration $\mathcal{F}M$ such that $G^{\mathcal{F}}(M) = \sum_{i \in J} G^{\mathcal{F}}(A)\sigma(\xi_i)$ with $\xi_i \in M$ and $\deg\sigma(\xi_i) = \alpha(i) \in I\!N^n$, then $M = \sum_{i \in J} A\xi_i$. In particular, if $G^{\mathcal{F}}(M)$ is finitely generated then so is M.
(ii) If M is finitely generated, then M has a $\succeq_{gr}$-filtration $\mathcal{F}M$ such that $G^{\mathcal{F}}(M)$ is finitely generated over $G^{\mathcal{F}}(A)$.

Proof (i) Since $G^{\mathcal{F}}(M) = \sum_{i \in J} G^{\mathcal{F}}(A)\sigma(\xi_i)$ with $\xi_i \in M$ and $\deg\sigma(\xi_i) = \alpha(i) \in I\!N^n$, we have

$$G^{\mathcal{F}}(M)_\alpha = \sum_{\substack{i \in J \\ \beta(i)+\alpha(i)=\alpha}} G^{\mathcal{F}}(A)_{\beta(i)}\sigma(\xi_i), \quad \alpha \in I\!N^n.$$

Thus, for any $m \in \mathcal{F}_\alpha M$, $m = \sum a_{\beta(i)}\xi_i + m'$, where $a_{\beta(i)} \in \mathcal{F}_{\beta(i)}A$ with $\beta(i) + \alpha(i) = \alpha$, $m' \in \mathcal{F}_{\alpha^*}M$. Similarly we have $m' = \sum a'_{\gamma(i)}\xi_i + m''$, where $a'_{\gamma(i)} \in \mathcal{F}_{\gamma(i)}A$ with $\gamma(i) \mid \alpha(i) = \alpha^*$, $m'' \in \mathcal{F}_{\alpha^{**}}M$. Since $\alpha \succ_{gr} \alpha^* \succ_{gr} \alpha^{**}$ and $\succeq_{gr}$ is a graded monomial ordering, after a finite number of repetition of the above procedure, we arrive at

$$m \in \sum_{i \in J} \left(\sum_{\substack{\gamma \in [0,\alpha] \\ \gamma(i)+\alpha(i)=\gamma}} \mathcal{F}_{\gamma(i)}A \right) \xi_i$$

and it follows that

$$\mathcal{F}_\alpha M = \sum_{i \in J} \left(\sum_{\substack{\gamma \in [0,\alpha] \\ \gamma(i)+\alpha(i)=\gamma}} \mathcal{F}_{\gamma(i)}A \right) \xi_i$$

because $\xi_i \in \mathcal{F}_{\alpha(i)}M$, $i \in J$. Hence $M = \sum_{i \in J} A\xi_i$.
(ii) Suppose $M = \sum_{i=1}^{s} A\xi_i$ and $\{\xi_1, ..., \xi_s\}$ is a *minimal* set of generators for M. Choose $\alpha(1), ..., \alpha(s) \in I\!N^n$ arbitrarily and set

$$\sum_{\substack{\gamma \in [0,\alpha] \\ \gamma(i)+\sigma(i)=\gamma}} \mathcal{F}_{\gamma(i)}A = \{0\}$$

if $\gamma = \alpha(i) + x$ has no solution for any $\gamma \in [0, \alpha]$. Then, it is easy to see that the family $\mathcal{F}M$ consisting of

$$\mathcal{F}_\alpha M = \sum_{i=1}^{s} \left(\sum_{\substack{\gamma \in [0,\alpha] \\ \gamma(i)+\alpha(i)=\gamma}} \mathcal{F}_{\gamma(i)} A \right) \xi_i, \quad \alpha \in I\!N^n,$$

is a $\succeq_{gr}$-filtration on M, where $\xi_i \in \mathcal{F}_{\alpha(i)}M - \mathcal{F}_{\alpha(i)*}M$, i.e., $\deg\xi_i = \alpha(i)$, $i = 1,...,s$. And by Lemma 2.2, it can be verified directly that $G^{\mathcal{F}}(M) = \oplus_{i=1}^{s} G^{\mathcal{F}}(A)\sigma(\xi_i)$ with

$$G^{\mathcal{F}}(M)_\alpha = \sum_{\substack{1 \le i \le s \\ \gamma(i)+\alpha(i)=\alpha}} G^{\mathcal{F}}(A)_{\gamma(i)}\sigma(\xi_i), \quad \alpha \in I\!N^n.$$

$\square$

Let M and N be $\succeq_{gr}$-filtered A-modules with $\succeq_{gr}$-filtrations $\mathcal{F}M$ and $\mathcal{F}N$, respectively. An A-module homomorphism $\varphi\colon M \to N$ is said to be a $\succeq_{gr}$-filtered homomorphism, if $\varphi(\mathcal{F}_\alpha M) \subset \mathcal{F}_\alpha N$ for all $\alpha \in I\!N^n$. A $\succeq_{gr}$-filtered homomorphism φ is said to be strict if

$$\varphi(\mathcal{F}_\alpha M) = \varphi(M) \cap \mathcal{F}_\alpha N, \quad \alpha \in I\!N^n.$$

If M is a $\succeq_{gr}$-filtered A-module with $\succeq_{gr}$-filtration $\mathcal{F}M$, and if $N \subset M$ is an A-submodule of M, then N has the $\succeq_{gr}$-filtration $\mathcal{F}N$ consisting of

$$\mathcal{F}_\alpha N = \mathcal{F}_\alpha M \cap N, \quad \alpha \in I\!N^n,$$

and the quotient A-module M/N has the $\succeq_{gr}$-filtration $\mathcal{F}(M/N)$ consisting of

$$\mathcal{F}_\alpha(M/N) = (\mathcal{F}_\alpha M + N)/N, \quad \alpha \in I\!N^n.$$

The $\succeq_{gr}$-filtrations $\mathcal{F}N$ and $\mathcal{F}(M/N)$ defined above are called the induced $\succeq_{gr}$-filtration on N and M/N, respectively. With respect to the induced filtration on N and M/N, the inclusion map $N \to M$ and the natural map $M \to M/N$ are strict $\succeq_{gr}$-filtered homomorphisms.

If $\varphi\colon M \to N$ is a $\succeq_{gr}$-filtered A-homomorphism, then φ induces naturally a $I\!N^n$-graded $G^{\mathcal{F}}(A)$-module homomorphism:

$$G^{\mathcal{F}}(\varphi)\colon \quad G^{\mathcal{F}}(M) = \bigoplus_{\alpha \in I\!N^n} G^{\mathcal{F}}(M)_\alpha \quad \longrightarrow \quad \bigoplus_{\alpha \in I\!N^n} G^{\mathcal{F}}(N)_\alpha = G^{\mathcal{F}}(N)$$

$$\sum \overline{m} \quad\quad\quad\quad \mapsto \quad\quad\quad\quad \sum \overline{\varphi(m)}$$

2.4. Proposition Let

$$(*) \quad\quad\quad\quad\quad\quad\quad\quad K \xrightarrow{\varphi} M \xrightarrow{\psi} N$$

be a sequence of $\succeq_{gr}$-filtered A-modules and $\succeq_{gr}$-filtered homomorphisms such that $\psi \circ \varphi = 0$. Then

$$G^{\mathcal{F}}(*) \qquad\qquad G^{\mathcal{F}}(K) \xrightarrow{G^{\mathcal{F}}(\varphi)} G^{\mathcal{F}}(M) \xrightarrow{G^{\mathcal{F}}(\psi)} G^{\mathcal{F}}(N)$$

is an exact sequence of $I\!N^n$-graded $G^{\mathcal{F}}(A)$-modules and $I\!N^n$-graded homomorphisms if and only if $(*)$ is exact and φ, ψ are strict.

Proof First suppose that $(*)$ is exact and φ, ψ are strict. If $G^{F}(\psi)(\overline{m}) = 0$ with $m \in \mathcal{F}_\alpha M - \mathcal{F}_{\alpha^*} M$, then $0 = \overline{\psi(m)} \in G^{\mathcal{F}}(N)_\alpha$, i.e., $\psi(m) \in \mathcal{F}_{\alpha^*} N \cap \psi(M) = \psi(\mathcal{F}_{\alpha^*} M)$. Thus, $\psi(m) = \psi(m')$ for some $m' \in \mathcal{F}_{\alpha^*} M$, and hence $m - m' \in \operatorname{Ker}\psi \cap \mathcal{F}_\alpha M = \varphi(K) \cap \mathcal{F}_\alpha M = \varphi(\mathcal{F}_\alpha K)$. Let $m - m' = \varphi(k)$ for some $k \in \mathcal{F}_\alpha K$. Then $\overline{m} = \overline{m - m'} = \overline{\varphi(k)} = G^{\mathcal{F}}(\varphi)(k)$. This shows that $\operatorname{Ker}G^{\mathcal{F}}(\psi) = G^{\mathcal{F}}(\varphi)(G^{\mathcal{F}}(K))$, i.e., the graded sequence is exact.

Conversely, suppose that the graded sequence $G^{\mathcal{F}}(*)$ is exact. To show the strictness of ψ, let $f \in \mathcal{F}_\alpha N \cap \psi(M)$ and $f \notin \mathcal{F}_{\alpha^*} N$. Then $f = \psi(m)$ for some $m \in \mathcal{F}_\beta M$ where $\beta \succeq_{gr} \alpha$. If $\beta = \alpha$, then $f = \psi(m) \in \psi(\mathcal{F}_\alpha M)$. If $\beta \succ_{gr} \alpha$, then since $f \in \mathcal{F}_\alpha N$, we have $G^{\mathcal{F}}(\psi)(\overline{m}) = \overline{\psi(m)} = 0$ in $G^{\mathcal{F}}(N)$. By the exactness, $\overline{m} = G^{\mathcal{F}}(\varphi)(\overline{k}) = \overline{\varphi(k)}$ for some $k \in \mathcal{F}_\beta K$. Put $m' = m - \varphi(k)$. Then $m' \in \mathcal{F}_{\beta^*} M$, and $\psi(m') = \psi(m - \varphi(k)) = \psi(m) = f$. Note that the chain

$$\beta \succ_{gr} \beta^* \succ_{gr} \beta^{**} \succ_{gr} \cdots \succ_{gr} \alpha$$

has finite length in $I\!N^n$. It follows that, after a finite number of repetition of the above procedure, we have $f = \psi(m_\alpha) \in \psi(\mathcal{F}_\alpha M)$. This shows that $\mathcal{F}_\alpha N \cap \psi(M) \subset \psi(\mathcal{F}_\alpha M)$, i.e, ψ is strict. A similar argument does reach the strictness of φ and the exactness of $(*)$. $\qquad\square$

2.5. Corollary Let $\varphi \colon M \to N$ be a $\succeq_{gr}$-filtered A-homomorphism. Then $G^{\mathcal{F}}(\varphi)$ is injective, respectively surjective, if and only if φ is injective, respectively surjective, and φ is strict.

3. General Case: gl.dim$A \leq n$

Let $A = k[a_1, ..., a_n]$ be an arbitrary quadric solvable polynomial algebra, and let $\mathcal{F}A$ be the $\succeq_{gr}$-filtration on A, as before. With the preparation made in §2, we proceed to show gl.dim$A \leq n$ in the present section.

First recall a well-known result concerning graded projective modules over a (semi)group-graded ring (e.g., [NVO]). Let G be an additive (semi)group and $S = \oplus_{g \in G} S_g$ a G-graded ring. A *graded free* S-module is a free S-module $T = \oplus_{i \in J} Se_i$ on the basis $\{e_i\}_{i \in J}$, which is also G-graded such that each e_i is

homogeneous, i.e, if $\deg(e_i) = g_i$, then $T = \oplus_{g \in G} T_g$ with $T_g = \oplus_{h_i + g_i = g} S_{h_i} e_i$.
For any graded S-module $M = \oplus_{g \in G} M_g$, there is a graded free S-module $T = \oplus_{g \in G} T_g$ and a graded surjective S-homomorrphism $\varphi \colon T \to M$. If T is a graded free S-module and P is a graded S-module such that $T = P \oplus Q$ for some graded S-module Q with the property that $T_g = P_g + Q_g$ for all $g \in G$, then P is called a *graded projective* S-module.

3.1. Proposition Let G be a (semi)group, S a G-graded ring and P a graded (left) S-module. The following are equivalent.
(i) P is a graded projective S-module.
(ii) Given any exact sequence of graded S-modules and graded S-homomorphisms $M \xrightarrow{\psi} N \to 0$, if $P \xrightarrow{\alpha} N$ is a graded S-homomorphism, then there exists a unique graded homomorphism $P \xrightarrow{\varphi} M$ making the following diagram commute:

$$
\begin{array}{ccccc}
 & & P & & \\
 & \varphi \swarrow & \downarrow \alpha & & \\
M & \xrightarrow{\psi} & N & \to & 0
\end{array}
$$

(iii) P is projective as an (ungraded) S-module.

$\square$

Return to modules over the quadric solvable polynomial algebra A. Let $L = \oplus_{i \in J} A e_i$ be a free A-module on the basis $\{e_i\}_{i \in J}$. In view of Lemma 2.2 and the proof of Proposition 2.3, chosen $\alpha(i) \in I\!N^n$ arbitrarily for $i \in J$, we may define a $\succeq_{gr}$-filtration $\mathcal{F}L$ on L:

$$
\mathcal{F}_\alpha L = \bigoplus_{i \in J} \left(\sum_{\substack{\gamma \in [0, \alpha] \\ \gamma(i) + \alpha(i) = \gamma}} \mathcal{F}_{\gamma(i)} A \right) e_i, \quad \alpha \in I\!N^n,
$$

where $[0, \ \alpha] = \{\gamma \in I\!N^n \mid \alpha \succeq_{gr} \gamma\}$ as defined before Lemma 2.2, and

$$
\sum_{\substack{\gamma \in [0, \alpha] \\ \gamma(i) + \alpha(i) = \gamma}} \mathcal{F}_{\gamma(i)} A = \{0\}
$$

if $\gamma = \alpha(i) + x$ has no solution in $I\!N^n$ for any $\gamma \in [0, \ \alpha]$.

3.2. Observation In the construction of $\mathcal{F}L$ made above, the following properties may be verified directly by using the monomial ordering $\succeq_{gr}$.
(i) For each $\alpha \in I\!N^n$ and each $i \in J$,

$$
\text{either} \sum_{\substack{\gamma \in [0, \alpha] \\ \gamma(i) + \alpha(i) = \gamma}} \mathcal{F}_{\gamma(i)} A = \{0\} \quad \text{or} \sum_{\substack{\gamma \in [0, \alpha] \\ \gamma(i) + \alpha(i) = \gamma}} \mathcal{F}_{\gamma(i)} A = \mathcal{F}_{\widetilde{\gamma}(i)} A
$$

where $\tilde{\gamma}(i) = \max\{\gamma(i) \in I\!N^n \mid \gamma(i) + \alpha(i) = \gamma$ for some $\gamma \in [0, \alpha]\}$.
(ii) For each $i \in J$, $e_i \in \mathcal{F}_{\alpha(i)}L - \mathcal{F}_{\alpha(i)*}L$, i.e., each e_i is of degree $\alpha(i)$.

3.3. Definition Write $\mathcal{F}L = \{\mathcal{F}_\alpha L; \ \alpha(i), \ i \in J\}_{\alpha \in I\!N^n}$ for the $\succeq_{gr}$-filtration on L as defined above. L is called a $\succeq_{gr}$-*filtered free* A-module with the $\succeq_{gr}$-filtration $\mathcal{F}L$.

3.4. Proposition With notation as above, the following holds.
(i) If L is a $\succeq_{gr}$-filtered free A-module with the $\succeq_{gr}$-filtration $\mathcal{F}L$, then $G^{\mathcal{F}}(L)$ is a $I\!N^n$-graded free $G^{\mathcal{F}}(A)$-module.
(ii) If L' is a $I\!N^n$-graded free $G^{\mathcal{F}}(A)$-module, then $L' \cong G^{\mathcal{F}}(L)$ for some $\succeq_{gr}$-filtered free A-module L.
(iii) If L is a $\succeq_{gr}$-filtered free A-module with the $\succeq_{gr}$-filtration $\mathcal{F}L$, N is a $\succeq_{gr}$-filtered A-module with $\succeq_{gr}$-filtration $\mathcal{F}N$ and $\varphi\colon G^{\mathcal{F}}(L) \to G^{\mathcal{F}}(N)$ is a graded surjection, then $\varphi = G^{\mathcal{F}}(\psi)$ for some strict $\succeq_{gr}$-filtered surjection $\psi\colon L \to N$.

Proof Let $\mathcal{F}L = \{\mathcal{F}_\alpha L; \ \alpha(i), \ i \in J\}_{\alpha \in I\!N^n}$ be the $\succeq_{gr}$-filtration on the free A-module $L = \oplus_{i \in J} Ae_i$. By Lemma 2.2, it can be verified directly that, for $\alpha \in I\!N^n$,

$$G^{\mathcal{F}}(L)_\alpha = \bigoplus_{\substack{i \in J \\ \beta(i)+\alpha(i)=\alpha}} G^{\mathcal{F}}(A)_{\beta(i)} \sigma(e_i),$$

where each $\sigma(e_i)$ is a homogeneous element of degree $\alpha(i)$ and $\{\sigma(e_i)\}_{i \in J}$ forms a free $G^{\mathcal{F}}(A)$-basis for $G^{\mathcal{F}}(L)$. This proves (i), and then (ii) follows immediately.
(iii) Let $L = \oplus_{i \in J} Ae_i$ with the $\succeq_{gr}$-filtration $\mathcal{F}L = \{\mathcal{F}_\alpha L; \ \alpha(i), \ i \in J\}_{\alpha \in I\!N^n}$. For each i, choose $x_i \in \mathcal{F}_{\alpha(i)}N$ such that $\varphi(\sigma(e_i)) = \bar{x}_i$, where $\bar{x}_i$ is the homogeneous element in $G^{\mathcal{F}}(N)_{\alpha(i)}$ represented by x_i. Now $\psi\colon L \to N$ may be constructed by putting

$$\psi\left(\sum a_i e_i\right) = \sum a_i x_i, \text{ where } \sum a_i e_i \in L.$$

Clearly, ψ is a $\succeq_{gr}$-filtered homomorphism and $G^{\mathcal{F}}(\psi) = \varphi$ since they agree on generators. By Corollary 2.5, ψ is a strict $\succeq_{gr}$-filtered surjection. □

3.5. Proposition Let P be a $\succeq_{gr}$-filtered A-module with $\succeq_{gr}$-filtration $\mathcal{F}P$. The following holds.
(i) If $G^{\mathcal{F}}(P)$ is a projective $G^{\mathcal{F}}(A)$-module, then P is a projective A-module.
(ii) If $G^{\mathcal{F}}(P)$ is a $I\!N^n$-graded free $G^{\mathcal{F}}(A)$-module, then P is a free A-module.

Proof (i) By Proposition 3.4, let $\varphi\colon G^{\mathcal{F}}(L) \to G^{\mathcal{F}}(P)$ be a graded surjection, where L is a $\succeq_{gr}$-filtered free A-module and hence $G^{\mathcal{F}}(L)$ is a graded free $G^{\mathcal{F}}(A)$-module. Again by Proposition 3.4, $\varphi = G^{\mathcal{F}}(\psi)$ for some strict $\succeq_{gr}$-filtered surjection $\psi\colon L \to P$. Let $K = \text{Ker}\psi$ and $\mathcal{F}K$ the $\succeq_{gr}$-filtration on K induced by $\mathcal{F}L$: $\mathcal{F}_\alpha K = K \cap \mathcal{F}_\alpha L$, $\alpha \in \mathbb{N}^n$. There is the short exact sequence

$$0 \to K \xrightarrow{\ell} L \xrightarrow{\psi} P \to 0$$

and it follows from Proposition 2.4 and Corollary 2.5 that the sequence

$$0 \to G^\mathcal{F}(K) \xrightarrow{G^\mathcal{F}(\ell)} G^\mathcal{F}(L) \xrightarrow{G^\mathcal{F}(\psi)} G^\mathcal{F}(P) \to 0$$

is exact. By Proposition 3.1, this sequence splits by graded $G^\mathcal{F}(A)$-homomorphisms. Consequently, $G^\mathcal{F}(L) = G^\mathcal{F}(P) \oplus G^\mathcal{F}(K)$ with $G^\mathcal{F}(L)_\alpha = G^\mathcal{F}(P)_\alpha \oplus G^\mathcal{F}(K)_\alpha$, $\alpha \in \mathbb{N}^n$, and there is a graded surjection γ: $G^\mathcal{F}(L) \to G^\mathcal{F}(K)$ such that $\gamma \circ G^\mathcal{F}(\ell) = 1_{G^\mathcal{F}(K)}$. By Proposition 3.4(iii), $\gamma = G^\mathcal{F}(\beta)$ for some strict $\succeq_{gr}$-filtered surjection β: $L \to K$. Note that $G^\mathcal{F}(\beta) \circ G^\mathcal{F}(\ell) = G^\mathcal{F}(\beta \circ \ell) = 1_{G^\mathcal{F}(k)}$. It follows from Corollory 2.5 that $\beta \circ \ell$ is an automorphism of K, and hence $L \cong K \oplus P$. This shows that P is projective.

(ii) Suppose $G^\mathcal{F}(P) = \oplus_{i \in J} G^\mathcal{F}(A)\sigma(\xi_i)$, where each $\xi_i \in P$ has degree $\alpha(i)$ and $\{\sigma(\xi_i)\}_{i \in J}$ is a $\mathbb{N}^n$-graded free basis for $G^\mathcal{F}(P)$ over $G^\mathcal{F}(A)$. Then, by Proposition 2.3, $P = \sum_{i \in J} A\xi_i$ with

$$\mathcal{F}_\alpha P = \bigoplus_{i \in J} \left(\sum_{\substack{\gamma \in [0, \alpha] \\ \gamma(i) + \alpha(i) = \gamma}} \mathcal{F}_{\gamma(i)} A \right) \xi_i, \quad \alpha \in \mathbb{N}^n.$$

We claim that $\{\xi_i\}_{i \in J}$ is a free basis for P over A. To see this, construct the $\succeq_{gr}$-filtered free A-module $L = \oplus_{i \in J} Ae_i$ with the $\succeq_{gr}$-filtration $\mathcal{F}L = \{\mathcal{F}_\alpha L; \alpha(i), i \in J\}_{\alpha \in \mathbb{N}^n}$ as before, such that each e_i has the same degree $\alpha(i)$ as ξ_i does. Then we have an exact sequence of $\succeq_{gr}$-filtered A-modules and strict $\succeq_{gr}$-filtered A-homomorphisms

$$0 \longrightarrow K \longrightarrow L \xrightarrow{\varphi} P \longrightarrow 0$$

where K has the $\succeq_{gr}$-filtration induced by $\mathcal{F}L$, and it follows from Proposition 2.4 that this sequence yields an exact sequence

$$0 \longrightarrow G^\mathcal{F}(K) \longrightarrow G^\mathcal{F}(L) \xrightarrow{G^\mathcal{F}(\varphi)} G^\mathcal{F}(P) \longrightarrow 0$$

However, $G^\mathcal{F}(\varphi)$ is an isomorphism. Hence $G^\mathcal{F}(K) = \{0\}$ and then $K = \{0\}$. This proves that φ is an isomorphism, or in other words, P is free. $\square$

3.6. Proposition Let M be a $\succeq_{gr}$-filtered A-module with $\succeq_{gr}$-filtration $\mathcal{F}M$, and let

(1) $$0 \to K' \to L'_n \to \cdots \to L'_0 \to G^\mathcal{F}(M) \to 0$$

be an exact sequence of $\mathbb{N}^n$-graded $G^\mathcal{F}(A)$-modules and graded homomorphisms, where the L'_i are graded free $G^\mathcal{F}(A)$-modules. The following holds.

(i) There exists a corresponding exact sequence of $\succeq_{gr}$-filtered A-modules and strict $\succeq_{gr}$-filtered homomorphisms

$$(2) \qquad\qquad 0 \to K \to L_n \to \cdots \to L_0 \to M \to 0$$

in which the L_i are $\succeq_{gr}$-filtered free A-modules. Moreover, we have the isomorphism of chain complexes

$$
\begin{array}{ccccccccccc}
0 \to & K' & \to & L'_n & \to & \cdots & \to & L'_0 & \to & G^{\mathcal{F}}(M) & \to 0\\
& \cong\downarrow & & \cong\downarrow & & & & \cong\downarrow & & =\downarrow & \\
0 \to & G^{\mathcal{F}}(K) & \to & G^{\mathcal{F}}(L_n) & \to & \cdots & \to & G^{\mathcal{F}}(L_0) & \to & G^{\mathcal{F}}(M) & \to 0
\end{array}
$$

(ii) If K' is a projective $G^{\mathcal{F}}(A)$-module, then K is a projective A-module; If K' is a $I\!N^n$-graded free $G^{\mathcal{F}}(A)$-module, then K is a free A-module.

(iii) If the modules in (1) are finitely generated over $G^{\mathcal{F}}(A)$, then the modules in (2) are finitely generated over A.

Proof (i) By Proposition 3.4, the homomorphism $L'_0 \to G^{\mathcal{F}}(M)$ in (1) has the form $G^{\mathcal{F}}(\beta)$ for some strict $\succeq_{gr}$-filtered surjection $\beta\colon L_0 \to M$, where $L'_0 \cong G^{\mathcal{F}}(L_0)$ and L_0 is a $\succeq_{gr}$-filtered free A-module. Let $K_0 = \mathrm{Ker}\beta$ with the $\succeq_{gr}$-filtration $\mathcal{F}K_0$ induced by $\mathcal{F}L_0$. Then we have the exact diagram of graded $G^{\mathcal{F}}(A)$-modules and graded homomorphisms

$$
\begin{array}{ccccccccc}
\cdots & \to & L'_2 & \to & L'_1 & \to & L'_0 & \to & G^{\mathcal{F}}(M) \to 0\\
& & & & \cong\downarrow & & =\downarrow & & \\
& & 0 & \to & G^{\mathcal{F}}(K_0) & \to & G^{\mathcal{F}}(L_0) & \to & G^{\mathcal{F}}(M) \to 0
\end{array}
$$

Note that the square involved in the above diagram commutes. Hence the homomorphism $L'_1 \to L'_0$ factors through $G^{\mathcal{F}}(K_0)$, i.e., there is the graded exact sequence $L'_1 \to G^{\mathcal{F}}(K_0) \to 0$. Starting with $G^{\mathcal{F}}(K_0)$, the foregoing construction can be repeated for finishing the proof of (i).

(ii) and (iii) follow immediately from Proposition 3.5 and Proposition 3.6, respectively. $\qquad\square$

We are ready to mention the finiteness of global dimension for A.

3.7. Theorem Let $A = k[a_1, ..., a_n]$ be an arbitrary quadric solvable polynomial algebra with the $\succeq_{gr}$-filtration $\mathcal{F}A$. Write p.dim for projective diemnsion and write gl.dim for global dimension. The following holds.

(i) If M is a $\succeq_{gr}$-filtered A-module with $\succeq_{gr}$-filtration $\mathcal{F}M$, then p.dim$M \leq$ p.dim$G^{\mathcal{F}}(M) \leq n$.

(ii) gl.dim$A \leq$ gl.dim$G^{\mathcal{F}}(A) = n$.

Proof Let $\mathcal{F}A$ be the $\succeq_{gr}$-filtration $\mathcal{F}A$ on A as defined in CH.V §6 and $G^{\mathcal{F}}(A)$ the associated $I\!N^n$-graded algebra of A. Then, by CH.V Proposition 6.1, $G^{\mathcal{F}}(A)$ is a homogeneous solvable polynomial algebra with respect to $\succeq_{gr}$. Hence, $G^{\mathcal{F}}(A)$ is an iterated skew polynomial algebra over the polynomial algebra $k[t]$ with all σ-derivations being 0, and consequently $G^{\mathcal{F}}(A)$ is an Auslander regular domain of global dimension n. Note that every A-module M has a $\succeq_{gr}$-filtration $\mathcal{F}M$. Therefore, (i) and (ii) follow from Proposition 3.6.

4. General Case: $K_0(A) \cong \mathbb{Z}$

We put the result as stated by the above title in this separate and final section just for emphasizing that we are backing to use the *standard filtration* again.

4.1. Theorem Let $A = k[a_1, ..., a_n]$ be an arbitrary quadric solvable polynomial algebra with its standard filtration FA. Let $G(A)$ and $\widetilde{A}$ be the associated graded algebra and Rees algebra of A with respect to FA, respectively. Then

$$\mathbb{Z} \cong K_0(A) = K_0(G(A)) = K_0(\widetilde{A}).$$

Proof By CH.IV Theorem 4.1–4.2, $G(A) = k[\sigma(a_1), ..., \sigma(a_n)]$ and $\widetilde{A} = k[\tilde{a}_1, ..., \tilde{a}_n, X]$ are quadric solvable polynomial algebras with respect to some $\succeq_{gr}$, respectively. It follows from Theorem 3.7 that $\mathrm{gl.dim} G(A) \leq n$ and $\mathrm{gl.dim} \widetilde{A} \leq n + 1$. Now, it follows from the K_0-part of Quillen's theorem ([Qui] Theorem 7) that

$$
\begin{array}{ccc}
 & K_0(F_0A) & \cong & K_0(A) \\
 & {\scriptstyle =\nearrow} & & \\
\mathbb{Z} \xrightarrow{\cong} K_0(k) & \xrightarrow{\cong} & K_0(G(A)_0) & \cong & K_0(G(A)) \\
 & {\scriptstyle =\searrow} & & \\
 & K_0(\widetilde{A}_0) & \cong & K_0(\widetilde{A})
\end{array}
$$

$\square$

Final remark Concerning the K-theory of iterated skew polynomial rings, the reader is referred to, e.g., [Art], [MR] and [Pas].

References

[ACG1] A. Assi, F.J. Castro-Jiménez and J.-M. Granger, How to calculate the slopes of a D-module, *Compsitio Math.*, 104(1996), 107–123.

[ACG2] A. Assi, F.J. Castro-Jiménez and J.-M. Granger, The Gröbner fan of an A_n-module, *J. Pure and Appl. Alg.*, 150(2000), 27–39.

[AdL] W.W. Adams and P. Loustaunau, *An Introduction to Gröbner Bases*, Graduate Studies in Math., Vol. 3, Amer. Math. Soc., 1994.

[AL] J. Apel and W. Lassner, An extension of Buchberger's algorithm and calculations in enveloping fields of Lie algebras, *J. Symbolic Comput.*, 6(1988), 361–370.

[Art] V.A. Artamonov, The quantum Serre problem, (Russian, English) *Russ. Math. Surv.*, 4(53)(1998), 657–730; translation from *Usp. Mat. Nauk*, 4(53)(1998), 3–76.

[AVV] M.J. Asensio, M. Van den bergh, and F. Van Oystaeyen, A new algebraic approach to microlocalization of filtered rings, *Trans. Amer. math. Soc.*, 2(316)(1989), 537–555.

[AW] R.A. Askey and J.A. Wilson, Some basic hypergeometric orthogonal polynomials that generalize Jacobi polynomials, *Mem. Amer. Math. Soc*, 318(1985).

[Bas] H. Bass, *Algebraic K-theory*, W. Benjamin, New York, 1968.

[Ben] G. Benkart, Down-up algebras and Witten's deformations of the universal enveloping algebra of sl_2, *Contemp. Math.*, 224(1999), 29–45.

[Ber] R. Berger, The quantum Poincaré-Birkhoff-Witt theorem, *Comm. Math. Physics*, 143(1992), 215–234.

[Berg] G. Bergman, The diamond lemma for ring theory, *Adv. Math.*, 29(1978), 178–218.

[Bern] I.N. Bernstein, Modules over a ring of differential operators, study of the fundamental solution of equations with constant coefficients, *Funct. Anal. Appl.*, 5(1971), 1–16(Russian); 89–101(English translation).

[Bj] J-E. Björk, *Rings of Differential Operators*, North-Holland Math. Library, Vol. 21, 1979.

[Bor] A. Borel et al, *Algebraic D-modules*, Perspectives in Mathematics, Vol.2, Academic Press, Boston,1987.

[BR] G. Benkart and T. Roby, Down-up algebras, *J. Alg.*, 209(1998), 305–344. Addendum: *J. Alg.*, 213(1999), 378.

[BW] T. Becker and V. Weispfenning, *Gröbner Bases*, Springer-Verlag, 1993.

[Cas] F. Castro, Calculs effectifs pour les idéaux d'opérateurs différentiels, in: *Géométrie Algébrique et Applications*, J. M. Aroca et al Eds., Travaux en Cours, Vol. 24, Hermann, Paris, 1987, 1–19.

[Cho] S.-C. Chou, *Mechanical Geometry Theorem Proving*, D. Reidel Publishing Company, Dordrecht, 1988.

[Chy] F. Chyzak, Holonomic systems and automatic proving of identities, *Research Report* 2371, Institute National de Recherche en Informatique et en Automatique, 1994.

[CLO'] D. Cox, J. Little and D. O'shea, *Ideals, Varieties, and Algorithms*, Springer-Verlag, 1992.

[CM] Paula A.A.B. Carvalho and Ian M. Musson, Down-up algebras and their representation theory, *J. Alg.*, 228(2000), 286–310.

[CS] F. Chyzak and B. Salvy, Noncommutative elimination in Ore algebras proves multivariate identities, *J. Symbolic Comput.*, 26(1998), 187–227.

[Dir] P. A. M. Dirac, On quantum algebra, *Proc. Camb. Phil. Soc.*, 23(1926), 412–418.

[Dix] J. Dixmier, *Enveloping Algebras*, North-Holland Math. Library, Vol.14, 1979.

[Eis] D. Eisenbud, *Commutative Algebra* with a view toward algebraic geometry, Springer-Verlag, 1995.

[FG] C.D. Feustel and E.L. Green, Gröbner, C codes, available anonymous ftp at http://www.math.vt.edu/people/green/index.html.

[Foa] D. Foata, Combinatoric des identities sur les polynômes orthogonaux, in: *Proc. Internat. Congress of Math.*, Warsaw, 1983, 1541–1553.

[Gal] A. Galligo, Some algorithmic questions on ideals of differential operators, *Proc. EUROCAL'85*, LNCS 204, 1985, 413–421.

[G-I1] T. Gateva-Ivanova, Algorithmic determination of Jacobson radical of monomial algebras, in: *Proc. EUROCAL'87*, LNCS, Vol.378, Springer-Verlag, 1989, 355–364.

[G-I2] T. Gateva-Ivanova, Global dimension of associative algebras, in: *Proc. AAECC-6*, LNCS, Vol.357, Springer-Verlag, 1989, 213–229.

[G-I3] T. Gateva-Ivanova, On the noetherianity of some associative finitely presented algebras, *J. Alg.*, 138(1991), 13–35.

[G-IL] T. Gateva-Ivanova and V. Latyshev, On recognizable properties of associative algebras, in: *Computational Aspects of Commutative Algebra*, from a special issue of the Journal of Symbolic Computation, L. Robbiano ed., Academic Press, 1989, 237–254.

[Gin] V. Ginsburg, Characteristic varieties and vanishing cycles, *Invent. Math.*, 84(1986), 327–402.

[Gol] E.S. Golod, Standard bases and homology, in: *Some Current Trends in Algebra*, Proceedings, Varna, 1986, LNM, Vol. 1352, Springer-Verlag, 1988, 88–95.

[Gr] E.L. Green, An introduction to noncommutative Gröbner bases, in: *Computational Algebra*, Proceedings of the fifth meeting of the Mid-Atlantic Algebra Conference, 1993, K.G. Fischer, P. Loustaunau, J. Shapiro, E.L. Green, and D. Farkas eds., Lecture Notes in Pure and Applied Mathematics, Vol. 151, Marcel Dekker, 1994, 167–190.

[GR] G. Gasper and M. Rahman, *Basic hypergeometric series*, Cambridge Univ. Press, 1990.

[Grö] W. Gröbner, *Algebraic Geometrie* I, II, Bibliographisches Institut, Mannheim, 1968, 1970.

[GZ] I.M. Gelfand and A.V. Zelevinskii, Algebraic and combinatorial aspects of the general theory of hypergeometric functions, *Functional Anal. Appl.*, 20(3)(1986), 183–197(English Translation).

[Hay] T. Hayashi, Q-analogues of Clifford and Weyl algebras-Spinor and oscillator representations of quantum enveloping algebras, *Comm. Math. Phys.*, 127(1990), 129–144.

[Jat] V.A. Jategaonkar, A multiplicative analogue of the Weyl algebra, *Comm. Alg.*, 12(1984), 1669–1688.

[JBS] A. Jannussis, G. Brodimas, and D. Sourlas, Remarks on the q-quantization, *Lett. Nuovo Cimento*, 30(1981), 123–127.

[Kas] M. Kashiwara, On the holonomic systems of linear differential equations II, *Invent. Math.*, 49(1978), 121–135.

[Kass] C. Kassel, *Quantum Groups*, Springer-Verlag, 1995.

[KL] G. Krause and T.H. Lenagan, *Growth of Algebras and Gelfand-Kirillov Dimension*, Research Notes in Math. 116, Pitman, London, 1985.

[KMP] E. Kirkman, I.M. Musson, and D.S. Passman, Noetherian down-up algebras, *Proc. Amer. Math. Soc.*, 127(1999), 2821–2827.

[K-RW] A. Kandri-Rody and V. Weispfenning, Non-commutative Gröbner bases in algebras of solvable type, *J. Symbolic Comput.*, 9(1990), 1–26.

[Kur] M.V. Kuryshkin, Opérateurs quantiques généralisés de création et d'annihilation, *Ann. Fond. L. de Broglie*, 5(1980), 111–125.

[KW] H. Kredel and V. Weispfenning, Computing dimension and independent sets for polynomial ideals, in: *Computational Aspects of Commutative Algebra*, from a special issue of the Journal of Symbolic Computation, L. Robbiano ed., Academic Press, 1989, 97–113.

[LC1] H. Li and L. Chen, On the recognition problem of the irreducibility for $A_1(k)$-modules and their characteristic varieties, in: *Interactions between ring theory and representation of algebras*, Proceedings of Euroconference 1998, F. van Oystaeyen Ed., Marcel Dekker, 2000, 277–285.

[LC2] H. Li and L. Chen, Multiplicity computation of modules over $k[x_1, ..., x_n]$ and an application to Weyl algebras, *Comm. Alg.*, 10(28)(2000), 4901–4917.

[Le1] L. Le Bruyn, two remarks on Witten's quantum sl_2 enveloping algebras, *Comm. Alg.*, 22(1994), 865–876.

[Le2] L. Le Bruyn, Conformal sl_2 enveloping algebras, *Comm. Alg.*, 23(1995), 1325–1362.

[LeS] L. Le Bruyn and S.P. Smith, Homogenized sl(2), *Proc. Amer. Math. Soc.*, 3(118)(1993), 725–730.

[Lev] T. Levasseur, Some properties of noncommutative regular graded rings, *Glasgow Math. J.*, 34(1992), 277–300.

[LeV] L. Le Bruyn and M. Van den Bergh, On quantum spaces of Lie algebras, *Proc. Amer. Math. Soc.*, 2(119)(1993), 407–414.

[Li1] H. Li, *Noncommutative Zariski Rings*, Thesis, Antwerp University, 1989.

[Li2] H. Li, Hilbert polynomial of modules over the homogeneous solvable polynomial algebras, *Comm. Alg.*, 5(27)(1999), 2375–2392.

[Li3] H. Li, Deformations of $U(sl_2)$ and their associated graded structures, *Unpublished note*, Bilkent, 2000.

[Li4] H. Li, Gröbner bases, Dickson systems, and algebras with divisors of zero, *to appear*.

[Li5] H. Li, Some computational problems in quadric solvable polynomial algebras, *to appear*.

[Li6] H. Li, Regularity and K_0-group of quadric solvable polynomial algebras, *to appear*.

[Lip1] L. Lipshitz, The diagonal of a D-finite power series is D-finite, *J. Alg.*, 113(1988), 373–378.

[Lip2] L. Lipshitz, D-finite power series, *J. Alg.*, 122(1989), 353–373.

[LVO1] H. Li and F. Van Oystaeyen, Zariskian Filtrations, *Comm. Alg.*, 17(1989), 2945–2970.

[LVO2] H. Li and F. van Oystaeyen, Global dimension and Auslander regularity of Rees rings, *Bull. Soc. Math. Belg.*, 43(1991), 59–87.

[LVO3] H. Li and F. Van Oystaeyen, Dehomogenization of gradings to Zariskian filtrations and applications to invertible ideals, *Proc. Amer. Math. Soc.*, 115(1)(1992), 1–11.

[LVO4] H. Li and F. Van Oystaeyen, *Zariskian Filtrations*, Monograph, Kluwer Academic Publishers, 1996.

[LVO5] H. Li and F. Van Oystaeyen, *A Primer of Algebraic Geometry – Constructive Computational Methods*, Pure and Applied Mathematics, Vol. 227, Marcel Dekker Inc. 2000.

[LVO6] H. Li and F. Van Oystaeyen, Elimination of variables in linear solvable polynomial algebras and ∂-holonomicity, *J. Alg.*, 234(2000), 101–127.

[LW] H. Li and Y. Wu, Filtered-graded transfer of Gröbner basis computation in solvable polynomial algebras, *Comm. Alg.*, 1(28)(2000), 15–32.

[LWZ] H. Li, Y. Wu and J. Zhang, Two applications of noncommutative Gröbner bases, *Ann. Univ. Ferrara - Sez. VII - Scienze Matematiche*, XLV(1999), 1–24.

[Man] Yu.I. Manin, *Quantum Groups and Noncommutative Geometry*, Les Publ. du Centre de Récherches Math., Universite de Montreal, 1988.

[Mor1] T. Mora, Gröbner basis for noncommutative polynomial rings, *Proc. AAECC3*, LNCS., 229(1986).

[Mor2] T. Mora, An introduction to commutative and noncommutative Gröbner bases, *Theoretical Computer Science*, 134(1994), 131–173.

[MP] J.C. McConnell and J.J. Pettit, Crossed products and multiplicative analogues of Weyl algebras, *J. London Math. Soc.*, (2)38(1988), 47–55.

[MR] J.C. McConnell and J.C. Robson, *Noncommutative Noetherian Rings*, John Wiley & Sons, 1987.

[NVO] C. Năstăsescu and F. Van Oystaeyen, *Graded ring theoey*, Math. Library 28, North Holland, Amsterdam, 1982.

[Oak1] T. Oaku, Gröbner bases and systems of linear partial differential equations–An introduction to computational algebraic analysis, *Sophia Kokyuroku in Mathematics* 38, Department of Mathematics, Sophia University, Tokyo, 1994.

[Oak2] T. Oaku, Algorithms for the b-function and D-modules associated with a polynomial, *J. Pure and Applied Algebra*, 117&118(1997), 495–518.

[Ore] O. Ore, Theory of noncommutative polynomials, *Ann. Math.*, 34(1933), 480–508.

[OT] T. Oaku and N. Takayama, An algorithm for de Rham cohomology groups of the complement of an affine variety via D-module computation, *J. Pure and Applied Algebra*, 139(1999), 201-233.

[Pas] D. Passman, *Infinite crossed products*, Acad. Press, San Diego, 1989.

[Pie] R.S. Pierce, *Associative Algebras*, Springer-Verlag, 1982.

[Qui] D. Quillen, Higher algebraic K-theory I, in *Algebraic K-theory I*: Higher K-theory, ed., H. Bass, Lecture Notes in Mathematics 341, Springer-Verlag, New York-Berlin, 1973, 85–147.

[Ros] A.L. Rosenberg, *Noncommutative Algebraic Geometry and Representations of Quantized Algebras*, Kluwer Academic Publishers, 1995.

[Rot] J.J. Rotman, *An Introduction to Homological Algebra*, Academic Press, 1979.

[Sm1] S.P. Smith, Quantum groups: an introduction and survey for ring theorists, in: *Noncommutative rings*, S. Montgomery and L. Small eds., MSRI Publ. 24(1992), Springer-Verlag, New York, 131–178.

[Sm2] S.P. Smith, A class of algebras similar to the enveloping algebra of sl(2), *Trans. Amer. Math. Soc.*, 332(1990), 285–314.

[Stan] P.R. Stanley, Differentiably finite power series, *Eur. J. Combin.*, 1(1980), 175–188.

[SST] M. Saito, B. Sturmfels and N. Takayama, *Gröbner Deformations of Hypergeometric Differential Equations*, Algorithms and Computation in Mathematics, 6, Springer-Verlag, 2000.

[Tak1] N. Takayama, An algorithm of constructing the integral of a module–an infinite dimensional analog of Gröbner basis, in: *Symbolic and Algebraic Computation*, Proceedings of ISSAC'90, Kyoto, ACM and addison-Wesley, 1990, 206–211.

[Tak2] N. Takayama, Gröbner basis, integration and transcendental functions, in: *Symbolic and Algebraic Computation*, Proceedings of ISSAC'90, Kyoto, ACM and Addison-Wesley, 1990, 152–156.

[Tak3] N. Takayama, An approach to the zero recognition problem by Buchberger algorithm, *J. Symbolic Comput.*, 14(1992), 265–282.

[Ufn] V.A. Ufnarovski, Calculations of growth and Hilbert series by computer, in: *Computational Algebra*, Proceedings of the fifth meeting of the Mid-Atlantic Algebra Conference, 1993, K.G. Fischer, P. Loustaunau, J. Shapiro, E.L. Green, and D. Farkas eds., Lecture Notes in Pure and Applied Mathematics, Vol. 151, Marcel Dekker, 1994, 247–255.

[Ufn1] V.A. Ufnarovski, Introduction to noncommutative Gröbner basis theory, in: *Gröbner Bases and Applications* (Linz, 1998), London Math. Soc. Lecture Notes Ser., 251, Cambridge Univ. Press, Cambridge, 1998, 259–280.

[Vdw] B.L. Van der Waerden, *Modern Algebra* II., Springer-Verlag, 1971.

[Wal] R. Wallisser, Rationale approximation der q-analogues der exponentialfunktion und Irrationalitätsaussagen für diese Funktion, *Arch. Math.*, 44(1985), 59–64.

[Wan] D. Wang, *Elimination Methods*, Texts and monograph in Symbolic Computation, Springer, 2000.

[Wit1] E. Witten, Gauge theories, vertex models, and quantum groups, *Nuclear Phys.*, B 330(1990), 285–346.

[Wit2] E. Witten, Quantization of Chern-Simons gauge theory with complex gauge group, *Comm. Math. Phys.*, 137(1991), 29–66.

[Wor] S.L. Woronowicz, Twisted SU(2) group. An example of a noncommutative differential calculus. *Publ. RIMS*, 23(1987), 117-181.

[Wey] H. Weyl, *Gruppentheorie und Quantenmechanik*, Hirzel, Leipzig, 1928 (English edition, Dover 1932).

[WZ] H.S. Wilf and D. Zeilberger, An algorithmic proof theory for hypergeometric (ordinary and "q") multisum/integral identities, *Invent. Math.*, 108(1992), 575–633.

[Zei1] D. Zeilberger, A holonomic system approach to special function identities, *J. Comput. Appl. Math.*, 32(1990), 321–368.

[Zei2] D. Zeilberger, The method of creative telescoping, *J. Symbolic Comput.*, 11(1991), 195–204.

[Zh] Zhang Jiangfeng, *The algebra of differential operators over a field of characteristic $p > 0$*, Thesis (in chinese), Xi'an Jiao-Tong University, 1999.

Index

Lecture Notes in Mathematics

For information about Vols. 1–1619
please contact your bookseller or Springer-Verlag

Recent Reprints and New Editions